W9-BDP-152

FINISHES

& FINISHING

TECHNIQUES

FINISHES

& FINISHING TECHNIQUES

**Professional
secrets for
simple and
beautiful
finishes from**
Fine Woodworking

The Woodworker's Library

The Taunton Press

Front cover photo by Vincent Laurence
Back cover photos by Alec Waters (top left), Aimé Fraser (top right),
Vincent Laurence (bottom)

Printed in the United States of America
10 9 8 7 6 5 4 3 2 1

Taunton
BOOKS & VIDEOS
for fellow enthusiasts

The Taunton Press, Inc., 63 South Main Street, PO Box 5506,
Newtown, CT 06470-5506
e-mail: tp@taunton.com

Distributed by Publishers Group West

Library of Congress Cataloging-in-Publication Data
Finishes & finishing techniques : professional secrets
for simple and beautiful finishes from Fine woodworking.
        p.   cm. — (The woodworker's library)
                ISBN 1-56158-302-2
    1. Wood finishing. I. Title: Finishes and finishing techniques. II. Series
                TT325.F5294   2000
        684'.084—dc21   99-053025
                        CIP

About Your Safety
Working with wood is inherently dangerous. Using hand or power tools improperly or ignoring
standard safety practices can lead to permanent injury or even death. Don't try to perform operations you
learn about here (or elsewhere) unless you're certain they are safe for you. If something about an
operation doesn't feel right, don't do it. Look for another way. We want you to enjoy the craft,
so please keep safety foremost in your mind whenever you're working with wood.

"If eyes were made for seeing, then beauty is its own excuse for being"

—RALPH WALDO EMERSON

CONTENTS

INTRODUCTION   3

ONE

SURFACE PREPARATION   5
Sanding in Stages   6
Fill the Grain for a Glass-Smooth Finish   12
Sealers: Secret for Finishing Success   17
Using Wood Putty   21

TWO

COLORING WOOD   27
Changing the Color of Wood   28
Dry-Brushing Wood Stains   33
Mix Your Own Oil Stains   38
For Vibrant Color, Use Wood Dyes   42
Glazes and Toners   50
Fuming with Ammonia   56
Using Wood Bleach   61

THREE

HAND FINISHING   67
Finishing Brushes   68
A Hand-Rubbed Oil Finish   74
Two-Day Lustrous Oil Finish   80
Padding On Shellac   85
Padding Lacquer   91
Making a Case for Varnish   96
Rubbing Out a Finish   102
Rejuvenating with Wax   107

FOUR

SPRAY FINISHING — 113

Taking the Spray-Finish Plunge — 114

Which Spray System Is Right for You? — 117

Build an HVLP Turbine with a Vacuum Motor — 122

Techniques for Blemish-Free Spraying — 128

FIVE

WATER-BASED FINISHES — 137

New Water-Based Finishes — 138

Using Waterborne Finishes — 146

SIX

SPECIAL TECHNIQUES — 155

Finish Cherry without Blotches — 156

Making Wood Look Old — 162

Creating an Antique Painted Finish — 167

Better Painted Furniture — 172

Repairing a Worn Finish without Refinishing — 178

Burning In Invisible Repairs — 181

Which Finishes Are Food Safe? — 189

A Case against Finishing — 195

ABOUT THE AUTHORS — 197

CREDITS — 198

EQUIVALENCE CHART — 199

INDEX — 200

INTRODUCTION

Every woodworker finds a good, simple way to finish projects and then sticks with it. Does Danish oil ring any bells? How about tung oil? Both are dead nuts simple to apply, essentially foolproof, and bring out a lovely, warm satin glow in just about any wood. Who could ask for more? Sometimes you need to: There will always come a project for which the old standby finish just won't do. Maybe it's a fancy dining table that begs for a high gloss finish or possibly a salad bowl that has to be absolutely nontoxic. Instead of trying to convince your wife that your formal dining table really would look better with a lovely, warm satin glow or that Danish oil is actually good to ingest, it's time to broaden your finishing horizons.

The need to know a range of finishes should be evident. First of all, the finish is to the furniture what the cover is to the book or the clothes to the man. Your furniture will be judged through its appearance, which is in large part determined by the finish (remember that you too will do much of that judging). Second, and perhaps more important, the finish protects and preserves the wood. A good finish is the first line of defense against the effects of time, wear, and damage.

Finishing is a craft unto itself, as rich and diverse with possibilities as furnituremaking. There are as many different finishes with pros and cons to pick from for any given project as there are projects. Needless to say, this can cause a fair amount of confusion. Many finishing techniques are difficult, and years of experience with a tablesaw do not aid finishing skills. It should not come as a surprise that the most common type of question sent to *Fine Woodworking* magazine concerns a finishing problem. In fact, finishing questions outnumber all the other types of questions by 2 to 1.

This book was put together with articles from *Fine Woodworking* that answer many of the questions and concerns woodworkers have asked over the years. You'll find articles that explain how to use pigments and dyes, that show you how to set up a spray system, and that offer a fair number of recipes and techniques for tried-and-true finishes. Each will give you just what you need to take a first step beyond Danish oil.

ONE

Surface Preparation

The one sure method to discover small, almost unnoticeable surface defects on a completed project is to apply a gloss film finish. What minute divots and scratches you couldn't see before now stare up at you, announcing their presence to all. It just doesn't seem fair—film finishes ought to cover defects, not highlight them. To add to the frustration, you can't begin sanding again until the finish has dried, which could be days away.

The best finishes don't start in the can, but on the furniture. Unless the surfaces have been properly prepared, you're just creating more work for yourself later on. Good surface preparation isn't just a matter of sanding down to the nth degree, but what type of surface you leave for the finish to interact with. One of the most important factors in the way a finish interacts with the wood is the wood's porosity, which determines how much finish penetrates the wood or stays on the surface. Woods vary greatly in this respect. Some, such as maple, aren't very porous and don't absorb much finish. Other woods, such as oak, are very porous and will drink up finish readily. Others yet, such as cherry and pine, can soak up finish and stain unevenly, leading to blotching. Some finishes won't work unless they can penetrate the wood, while for others it's the opposite case.

This chapter offers several methods for attaining the right kind of surface for a particular finish, not only in respect to smoothness and porosity but also to eliminating contaminants such as silicone, oil, and stearates that will prevent a finish from flowing out or adhering to the wood. These are important steps in all finishing—and worth the trouble. And when you read the instructions on the side of a can of finish, think of the advice on how the surface should be smooth and free of contaminants as a mental health warning. Stripping off the first coat of finish to redo the surface will certainly raise your blood pressure.

SANDING IN STAGES

by Gary Straub

Sanding is just as critical to the ultimate success of a piece of furniture as its design or joinery. Here, the author sands the frame of a frame-and-panel door, taking care of the tenoned rail first and then sanding the mortised stiles, thereby eliminating any stray scratch marks caused by sanding across the joint line. This process is repeated with each subsequent grit.

Everything from sharp stones to sharkskin has been used to smooth wood. Today there's a seemingly endless array of sanding tools, aids and abrasives available, all designed to make our work faster, easier and better. If we look back at the methods used to smooth wood, we should appreciate the ease with which we can produce results far superior to those of our predecessors. Even so, most woodworkers still dread sanding.

That's too bad because sanding is one of the most important aspects of producing a fine piece. No matter how much time and care go into the making of a piece, its

overall beauty is in large measure determined by how well it's been sanded. Although some finish representatives will tell you differently, no finish can cover up a mediocre sanding job.

Sanding doesn't have to be sheer drudgery, however, if you break the job down into its various stages and integrate the smoothing process with the construction of a piece of furniture. Before I've even ripped a board to width or crosscut it to length, I've beltsanded it to 100-grit. I remove all flaws with this preliminary sanding so that the only reason for further sanding is to remove the scratches created by the previous coarser grit. By the time I glue-up, everything's been sanded to 150-grit, which makes post-assembly sanding a breeze.

The result of this division of labor is a better sanding job, less tedium and a finer finished piece. The sanding system I've developed over the past twenty years of furnituremaking takes advantage of a wide range of abrasive materials. But before I explain my techniques, let's look at what's available today.

The materials

Sandpaper was invented when someone figured out how to glue screened particles of glass or sand onto a paper backing. Today, of course, true sandpaper and glasspaper are practically unavailable. They have been replaced by papers that use much harder and sharper minerals, both natural and synthetic. New abrasive materials, more sophisticated screening methods and superior papers and glues have transformed the ways we smooth wood. Not long ago, abrasives came in grits from 12 to 600; now they go into the thousands (see the photo above). As if that weren't enough, we also have steel wool, abrasive cloths, pads, powders, liquids and pastes. Knowing what to use has become a challenge.

Abrasives

In ascending order of hardness, the materials used for coated abrasives are glass, silica sand, garnet, aluminum oxide, silicon carbide and zirconia alumina. The abrasive is applied to a backing as an open or closed coat. A closed coat means there is complete coverage while an open coat has 40% to 60% coverage. Closed-coated abrasives are more aggressive but clog easier. Open-coated abrasives are less aggressive but don't clog as easily. Most wood sanding is best done with open-coated paper, but some very hard woods can be sanded with closed coat. Wet sanding can be done with closed-coated paper (the liquid keeps the abrasive from becoming clogged).

Backing materials

Backing materials come in weights from A to X, with A-wt. being the lightest. I use mostly A-wt., or finish paper, and C-wt.,

Paper- and cloth-backed abrasives are available in a huge range of grits, which are bonded with a variety of adhesives to backings of widely differing weights. They are all still called "sandpaper" even though none are made with sand.

cabinet or production paper. A-wt. is very flexible for hand- and finish-sanding. C-wt. is heavier but still fairly flexible for machine sanding. Discs are often E-wt. paper, cloth-backed sheets are J-wt. and sanding belts are usually X-wt. cloth.

Bonding agents

The abrasives can be bonded to the backing material with several different glues: hide glue for its flexibility, resins for their strength, or a combination of both. The grains may be electrostatically arranged, and often another coat of resin is added to maintain orientation. This is resin over resin and is used on better sanding belts and in other applications where strength of bond is more important than flexibility.

Reading the paper

Each company has its own method of displaying product information on the back of the sheets of sand-paper (see the top photo on p. 8). The type of abrasive is often written out fully (aluminum oxide or garnet, for example). The grit is displayed by a number, sometimes preceded by a letter, such as P100, and the coating may either be written out fully or abbreviated (Open Coat or OP). The backing weight may be shown as A or A-wt., or combined with either the grit designation (120A) or the information on coating density (AOP).

Choosing a paper

Generally, I use aluminum oxide papers with my portable sanding machines and switch to garnet for hand-sanding. Aluminum oxide lasts longer than garnet because it's a lot harder and so is more suited to machine sanding. It doesn't break down, however, so

For stripping paint or varnish and for cleaning metal or going over a finish, steel wool and abrasive pads work better than sandpaper.

the sharp edges will become dull. The combination of a dull belt and the speed of the machine (especially a belt sander) can severely burnish the wood, which could affect how it finishes. Dull belts should be replaced. Garnet continuously breaks down, exposing fresh sharp edges, but because it's softer than aluminum oxide, I use it only for hand-sanding.

I sand from 80 to 220 using aluminum oxide and garnet papers but use silicon carbide for grits 240 to 320. I also use coated abrasives on occasion. These papers are often silicon carbide, coated with a material, such as zinc stearate, that prevents the papers from clogging. I've found them helpful in sanding oily or resinous woods, but (contrary to what the manufacturers will tell you) there's a possibility of the residue contaminating the finish.

Non-paper abrasives

In addition to sandpaper, I also use 3M's Scotch-Brite or Norton's Bear-Tex nylon pads and steel wool. The pads are made of abrasive-coated fibrous nylon. They're very flexible, they last much longer than steel wool and they come in different grades, from coarse to ultra-fine. They're also good for wet sanding because they're unaffected by water, oil or solvents.

I use 00, 000 and 0000 (progressively finer) steel wool for finish work and sometimes use the coarser grades for stripping or for routine chores like metal cleaning. I like the steel wool for finish work because it cuts better than the abrasive pads, and the steel wool burnishes the wood slightly, which gives it a better sheen.

The method

The sanding process needn't be the hassle that we often make it. I've found that sanding as I go produces better results and takes much of the monotony out of the work. I

first plane or re-plane all lumber for a piece before I start. I keep my blades very sharp, and I never take more than $1/32$ in. per pass. On smaller pieces, I use a handplane. Lumber planed at either the lumberyard or mill is very crudely done and of poor consistency. Trying to sand mill-planed lumber flat is a waste of time.

Using machines

After planing all the lumber to a consistent thickness, I sand each piece with a portable belt sander and a 100-grit belt. This sanding is crucial because this is when I remove any flaws. It's tempting to decide that you've sanded enough and that the next grit will take care of the rest. This is never true. If you remove all the flaws on the first sanding, subsequent sandings need only remove the scratches left by the previous grit, thereby saving time overall.

Using a portable belt sander takes some practice because it's quite easy to remove far more wood than you want. Most sanders are not well-balanced, usually weighing more on one side or the other, or more toward the front or back. To compensate for this, you must exert a slight pressure opposite the weight, striving to maintain total contact with the surface. At the same time, you must keep the pressure equal in all directions. Leaning the machine to one side or the other will create long gouges. Applying too much pressure either to the front or back will cause dips.

The proper technique is to move the balanced machine back and forth slowly, with the grain, reaching comfortably but not stretching. Don't move the machine directly to the side but rather let it drift to the side as you go back and forth. Moving it sideways will cause zig-zag dips that usually remain hidden until the first coat of finish is applied.

I change belts as soon as I feel myself applying more pressure to get the belt to bite. Increasing pressure as belts dull is a primary cause of a poor sanding job. Unfortunately, the high cost of belts stimulate this bad habit. Cleaning the belt with a crepe-rubber bar belt-cleaning stick will stretch the life of your belts, but when they're dull, they're dull.

Having a brand new belt clog up with resin or glue can be very frustrating. I've had some success cleaning belts with a brass-bristled brush and in worse cases, using pitch cleaner with the brass brush. I do save all my used belts because they're still good enough for lathe work and for hand-sanding curved surfaces. I like Hermes aluminum oxide, resin-over-resin, open-coated belts. They're good belts at a fair price.

After sanding all flaws out of the lumber, I cut all stock to size, joint all the edges (finishing with a handplane), make all my joints and then dry-assemble. Next I glue up any wide panels such as tabletops. While they're drying, I sand the rest of the flat parts with a belt sander using 120-grit. All the parts that can't be sanded with a machine, I'll hand-sand with the same grit. Before sanding and between each grit, I brush each piece thoroughly to remove any residual grit—the cause of those mysterious scratches that often appear.

This sanding goes very fast, but you must be careful, especially on the edges. The only object of this sanding is to remove the previous sanding scratches because I've already removed all defects with the initial sanding. I then check for any dings that may have occurred while cutting, and if there are any, I'll put drops of water on them to raise the fibers. By this time, any wide-panel glue-ups are dry enough to remove the clamps. I use an old plane blade to remove excess glue before it dries completely; otherwise, it will lock moisture into the joints, causing problems later on.

Next, I handplane any irregularities in the glued-up surface because it's just not possible to make a large panel flat with a portable sander. Once I get the surface satisfactorily flat, I sand it with the belt sander using a 120-grit belt. I sand the back first so that I don't take a chance on scratching the top when I turn it over. I then do any decorative routing, inlays or carving, and I plane or hand-sand the panel again with 120-grit paper. Now all the pieces are made and sanded to 120-grit, which is fairly smooth.

Now I change to a half-sheet orbital sander and 150-grit aluminum oxide paper (I like Diamond Grit paper, made by the Carborundum Abrasives Co.), and then I go over all the flat surfaces before assembly.

This makes problem areas—such as joints where the grain goes in different directions—much easier to deal with after assembly. I then dry-assemble the piece to check for any variation in wood thickness at the joints. Sanding these flush now makes post-assembly sanding much easier and pleasant.

When everything looks and fits right, I glue up. Because there's no turning back now, I make sure I'm satisfied that all is ready. I use glue sparingly, so there is minimum squeeze-out (but I make sure there's a little, so I know the joint isn't starved). While waiting for the glue to set, I sand any wide panels to the same 150-grit.

A good orbital sander does an excellent job of sanding, removing wood quickly while maintaining flatness. I use a Porter-Cable 505 half-sheet sander and a Makita quarter-sheet sander. I always use the largest sander that will do the job, usually the half-sheet machine. I move the machine back and forth slowly with the grain, letting the machine do the work. I apply only enough pressure to maintain control. The quickest way to ruin both furniture and machine is to apply a lot of pressure. By applying just a little more pressure on the back of the sander on the forward stroke and on the front on the return stroke, I have more control and the machine performs better.

Moving slowly is key to minimizing swirl marks because it gives the paper a chance to erase them. Just as with the belt sander, I shift sideways slowly as I'm moving back and forth to avoid creating any swirl marks, and I brush my work often to prevent pieces of grit from getting caught under the pad. On very large panels, I sand one area at a time so that I don't forget where I've been.

The next step depends on the finish I'm using. I put oil on most of my work (except tabletops) because I like the way it allows the texture of the wood to be seen and felt. When finishing with oil, I stop at 150-grit. Oil is a penetrating finish, and the finer you sand, the less penetration you obtain. I apply the first coat (which does the most penetrating) before I go to finer abrasives.

For items requiring more protection, I use a surface finish such as varnish or lacquer. When I'm putting on a varnish finish, I continue machine sanding to 220-grit and for lacquer to 320-grit.

Always sand with a cushion to keep your surface flat. Cork blocks, cork-faced blocks and rubber sanding pads all will work. The rubber Tadpoles (right, background) allow sanding of concave and convex moldings, and the finger rubbers protect the fingers while providing a good grip on the sandpaper. The solid wood block is useful as a backing for the nylon abrasive pads.

Polishes further refine the finish. They include pumice and rottenstone as well as modern ultra-fine automotive products. In either case, felt is the best applicator.

Hand-sanding

Regardless of the finish, I always hand-sand all the pieces (except bottoms, backs and other parts that will not show) with the same grit that I used on the last machine sanding. This removes any remaining swirl marks and provides a good opportunity to examine every inch of the work.

Hand-sanding is labor-intensive, but it's also the most rewarding part of sanding. Using machines requires good balance and steady hands, but handwork lets you feel what you're doing. You must learn to detect slight imperfections with your hands to judge whether a curve is fair or an edge consistent.

When sanding flat surfaces by hand, you must use a sanding block to keep the surface flat (see the photo above). I prefer a solid-cork block, but I've also made sanding blocks by gluing bulletin-board cork (obtainable at most hardware stores) to a block of wood. Cork is firm enough to keep the paper flat and resilient enough not to destroy the paper. Some prefer felt- or rubber-faced sanding blocks. What's important is that you not use a block of wood alone, or it will quickly destroy the paper. The block I use takes a quarter sheet of paper.

I apply firm pressure to the block, stroking back and forth, carefully following the curves of the grain. I'm very careful with edges and corners, taking care not to round them off or taper them. If they're square, I try to keep them sharp for now. For miters, I hold the block at the same angle as the joint and sand up to the intersection from each direction. I deal with right-angle joints by sanding the tenoned section first. Then, when sanding the mortised section, I can remove any stray scratches.

Overexertion quickly leads to a hurry-up and get-it-done attitude. I take my time as if I were cutting feather-thin dovetails, sanding a small portion at a time and stopping often to brush away any loose grits. I check

my progress frequently, using my bare hand to tell me where I need to sand a little more. When I finish one section, I dust thoroughly, wipe with a soft clean cloth and then feel the surface again, making sure it's right.

The last step is to eliminate any sharp edges that I've left. Using the finest grit I've sanded with to this point, I go over the edges by hand, without a pad. I twist my hand slightly as I'm moving forward, which softens the edge more quickly than if I kept my hand fixed, and it prevents the edge from getting stuck in a groove in the paper. A very light touch will produce a corner that cannot be duplicated by any machine. A little more pressure will yield a $1/16$ in. radius in no time.

Sanding irregular surfaces

Sanding curved pieces is much the same as sanding flat surfaces except you have to begin hand-sanding right from the start. A flexible sanding block is important; I use rubber sanding pads, varying in firmness. Their flexibility allows them to bend to fit most curves.

For smaller curves and for sanding on the lathe, I tear a strip of whatever size I need from a used sanding belt I've saved. The heavy cloth back of the belt is pliable enough to fit the curve yet firm enough to maintain the shape. For small concave shapes, such as on moldings, I cut a piece of dowel that fits the groove and wrap it with A-wt. paper.

I also use rubber Tadpole Contour Sanding Grips (available from many mail-order woodworking catalogs). They come in various diameters, both concave and convex, and the flat grip section is shaped at the top to allow sanding in tight places. They come in sets, some include flexible sanding pads. They've made life easier, and they're very inexpensive.

Carve as smoothly and crisply as possible, so only minimal sanding is required. Carvings present the most difficulty because any sanding will alter the character of a carving. If it's a geometric carving or a large in-the-round carving, sanding with A-wt. paper works well. For heavy carvings, without fine detail, steel wool or abrasive pads conform well to irregular shapes. When I do a lot of sanding with my fingers, I wear finger rubbers (available at most office supply stores). They're made for office workers to flip easily through papers, but they're also perfect for protecting fingers, and giving a good grip on the sandpaper.

For highly detailed carvings, I use a stiff nylon-bristled brush—shaped like a toothbrush—and a slurry of powdered pumice and mineral spirits. Pumice is made from a type of lava and has been used through the ages as an abrasive both in the solid and crushed form. Powdered pumice is graded like steel wool except in F's instead of 0's. I use the finest grit that will work.

Smoothing the finish

After I'm satisfied that everything is smooth and ready for the finish, I wipe everything down with a soft rag dampened in mineral spirits. This serves three purposes: first, it cleans any contaminants that may have gotten on the wood, especially oils from my hands or drops of sweat from working on a hot summer day. It also gives me an idea how the piece will look finished and reveals any remaining imperfections. These are far easier to deal with now than after applying a finish.

The smoothing process isn't over when the finish goes on. Each coat must be abraded slightly before the next is applied either to ensure adhesion, as with varnish to remove dust specks in lacquer, or to finish the smoothing of an oil finish. I sand varnished and lacquered surfaces with 320-grit silicon carbide paper, often with water on varnish. But for my oil finish, I use steel wool starting with 00 and changing to the next finer grade with each coat. I prefer steel wool to the nylon abrasive pads because it not only smooths the surface by abrasion but also gently burnishes the oil-filled wood, creating a higher luster and a smoother feel.

The final coat of finish must also be smoothed or polished. A slurry of rottenstone (a very finely powdered mineral) mixed either with water or paraffin oil makes an excellent polish. Mixed with water, it gives a higher polish; mixed with oil, it gives a more satin finish.

Felt is the best material for the final rubbing. Felt blocks that look like sanding blocks are available commercially, but you can also make your own. The best felt comes from old felt hats that you might find in your father's attic or in used-clothing stores. The texture of that felt is very uniform, and it's stiff enough to use without a block for curved and carved parts. I just dip the felt in the slurry and rub with the grain. I rub with the felt by itself (no rottenstone) for oil finishes because I'm able to get the luster I want without abrasives. I rub harder and longer, though, because there's no danger of cutting through, now that the finish has become part of the wood.

There are many polishes for wood today that surpass rottenstone, so rottenstone is fading into history. Most finish companies either make a polish for their finish or recommend one. Also there are many automobile polishes that give excellent results on varnished or lacquered finishes. In fact, there are so many polishes available today that it's difficult to keep track of them all. I've been happy with Meguiar's Mirror Glaze, a brand I find at the local auto parts' store. It comes in varying degrees of abrasiveness. One caveat: Be careful when using polishes on wood whose grain has not been filled. The residue of many polishes will fill the grain and dry to a very unnatural color, which is extremely difficult to remove.

The last step is to remove any remaining polish with a very soft cloth. Cotton diapers are excellent but in short supply in this disposable society. Lint-free polishing cloths are available from finish suppliers or auto parts' stores. Wipe your piece down, and step back to admire a job well-done.

SOURCES OF SUPPLY

Sanding (and other abrasive) supplies are available from many general woodworking catalogs. There are also a number of companies whose specialty is abrasive products. Here are two that the author buys from.

Pyramid Products Co.
(800) 747-3600

Skates Belting,
(800) 821-5041

Klingspor Corp.
(800) 228-0000

FILL THE GRAIN FOR A GLASS-SMOOTH FINISH

By Chris A. Minick

Grain fillers are essential for getting glass-smooth finishes in open-pore woods. Here Chris Minick applies thinned water-based filler to a small area of a mahogany tabletop. Using a disposable brush, he works the bristles in a circle to pack the pores. Next he'll squeegee off excess filler before it dries. He masked off the table's edges to avoid having to scrape filler from the routed profile.

Woods like mahogany, ash, walnut and oak, which have large pores, give a natural open-grained appearance to furniture. But to get a glass-smooth surface on these woods, you have to fill the pores with a grain filler before applying the finish. Tight-grained hardwoods, like maple and most softwoods, usually don't require grain filling.

You need only a few tools to use grain fillers (see the photo on the facing page), and grain-filling is pretty straightforward: Thin and tint the filler, prepare the surface, brush on the filler and pack the pores, remove the excess before it hardens, sand to the wood once the filler is dry, and clean off any residue. But though the process is straightforward, filling grain takes time, is messy and is generally not much fun. Even so, the results are well worth the effort, as the left side of the butternut board shows in the photo below.

Oil-based and water-based options
Don't confuse grain fillers with the wood putty used to fill nail holes. Grain filler, also called paste-wood filler or pore filler, is a thick clay-like mixture of solvent, resin binders and finely ground minerals, often called silex. Fillers come in oil-based formulations, like Behlen's Pore-O-Pac (available from Woodcraft Supply, 210 Wood County Industrial Park, P.O. Box 1686, Parkersburg, WV 26102; 800-225-1153) or in water-based formulations, like Hydrocote's Fast Dry (available from Highland Hardware, 1045 N. Highland Ave. N.E., Atlanta, GA 30306; 800-241-6748). Both varieties can be purchased as a thick paste that must be thinned before use, or in a pre-thinned,

ready-to-use consistency. Even though oil-based grain fillers have been around longer, I prefer water-based fillers because they work easier, dry faster and are easier to clean up. In addition, water-based fillers, once completely dry, are compatible with virtually all finishes.

Tinting the filler
Pore fillers come in a variety of wood tones, so you can match your project. They also come in off-white and in a neutral color, which can be custom-tinted in your shop. The choice of tint is a matter of taste. You may want a light, unobtrusive filler color on oak, or you may want to contrast the grain by using a dark filler. I almost always go for a darker filler because I like to bring out grain patterns. Similarly, you can pick up highlights in the wood—reds in mahogany or maroon in walnut, for example. I usually stick with earth-tone pigments, such as burnt umber (chocolate-like color), ochre (yellowish), burnt sienna (reddish) and lamp black. To my eyes, colors that are bright and bold look artificial on wood.

If you decide not to tint your off-white or neutral oil-based filler, be aware that the binders in the mixture will likely cause the filler to yellow or darken with age. This is not a problem if you use a water-based grain filler. Pigmented universal tinting colors (UTCs), available from most large paint stores, and dry fresco powders work well at coloring water-based and oil-based fillers. Japan colors (pigments ground in a varnish base), artist's oil colors and the pigment sludge found on the bottoms of oil-based stain cans are only useful for tinting oil-

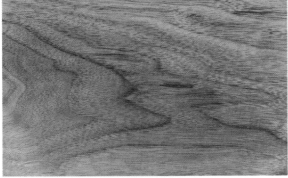

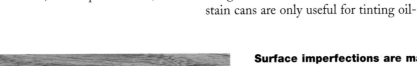

Surface imperfections are magnified by pore filler, so it's critical that the wood be properly scraped and sanded beforehand. The rust-colored filler used on this piece of butternut reveals even small surface blemishes. It's best to level off scratches and low spots using regular wood putty before applying pore filler.

based fillers. For more on tinting, see the photo and story below. In any case, make sure your coloring medium is a pigment. Transparent dye stains will not adequately color the quartz particles found in most grain fillers.

Preparing the surface

Sloppy sanding and pore fillers don't mix. That's why I usually power-sand the wood with a random-orbit sander through 180-grit sandpaper. Then I hand-sand with 220-grit to remove pesky swirl marks. Likewise, tearouts, gouges or other defects must be puttied and sanded flat before applying the filler. A poorly prepared sur-

face will be magnified a hundred fold once the blemishes are packed with pore filler (see the photo on p. 13).

Pore fillers tend to seal the wood surface, which makes staining after filling difficult. If you plan to stain the wood, do it before you fill the grain. I like to use water-based dye stains under the filler because the inevitable sand-throughs are easily repaired by reapplication of the same strength dye stain. Once the stain is dry, you should seal it (I prefer shellac or vinyl sealer). There are three reasons for sealing: First, the sealer protects the stain layer from scratches during the filling process. Second, sealing before filling eliminates an undesirable

Working with oil-based filler

by Andy Charron

Before you apply oil-based paste filler, you need to tint it to the right color for your project. The filler not only plugs up pores, but it will actually stain the immediate surrounding areas of the wood as well. Because the silex in the filler does not accept stain, you cannot readily change filler color once it's dry. However, the color can be adjusted beforehand by blending different fillers together or by adding pigments to neutral filler as described in the article. You can achieve the wood tone you want through trial and error (see the photo at right).

Besides needing tinting, oil-based filler usu-

ally has to be thinned as well. If the filler is too thick, you'll need lots of elbow grease to brush it on and to rub off the excess. If it's too thin, it will be easy to apply and clean up, but it won't fill up large pores well and it will take longer to dry. Once you've thinned the filler to a creamy consistency, apply it in stages over small areas. I like to use a stiff bristle brush to work the filler into the wood, appying it with the grain first and then going back over it perpendicular to the grain.

Immediately after you've filled the grain (while the wood is wet), remove all the excess by scraping the surface at a 45° angle. One filler manufacturer recommends using a piece of plastic, like a credit card,

for this. I've found a thin ripping of scrap works well. I cut one end of the scraper to a point so I can get filler out of hard-to-reach places, like inside corners.

Once you've scraped the surface clean, allow the residual filler to dry until it takes on a flat,

Start with neutral grain filler and add pigments, such as universal tinting colors (UTCs), to get the color you like.

crusty look (usually 5 to 10 minutes). At this point, start rubbing with a piece of burlap, which is rough enough to remove filler, yet porous enough to not get clogged. When the rag begins to weigh down with excess material, shake it out and it will be ready to wipe some more. Finally, polish the wood with a soft cotton cloth. Although one filler company recommends light sanding after you wipe the filler off, I've found that a good firm rubbing with a clean rag will shine the wood to a perfectly smooth, flat surface.

smudging effect that commonly occurs. Third, because sealers smooth out surfaces, they allow easy removal of excess filler.

Applying the filler and removing the excess

Once thinned to the consistency of heavy latex paint, pore filler is ready to apply. Paint on a fairly thick coat of filler (see the photo on p. 12), and then pack the filler into the pores using a forceful circular motion of the brush. (This is why I like to use disposable brushes.) Stir the filler frequently because filler particles are heavy and rapidly settle to the bottom of the can.

Remove excess filler from the surface with a stiff rubber squee-gee (available from a glass-cleaning supply store) for water-based filler or a plastic putty knife for oil-based filler. Pulling the squee-gee or pushing the putty knife diagonally across the grain minimizes the chance of removing the filler from the just-packed pores (see the top photo above). If you're using oil-based filler, use coarse burlap rags to clean residual filler off the wood before it dries. The more filler you remove now, the less sanding later.

Getting a feel for the proper drying time takes practice. Generally, you can begin removing an oil-based filler when the surface starts to look dull or hazy. A light sprinkling of mineral spirits over the filler

Sand down to the wood once the filler is dry. The author uses 120- and then 220-grit paper to produce the fine powder shown. If the paper starts to gum up, it means the filler is not quite dry.

will slow down the drying and allow a bit more working time. But water-based fillers dry so rapidly that if you wait for them to haze over, it's too late. Instead, work on small patches at a time and immediately squeegee the excess filler from the surface as soon as pore packing is complete. Because the filler won't leave lap marks, you don't have to fill the entire surface at once. But sprinkling water on hardened water-based filler is no help. If you wait too long to squeegee, you'll have to sand off the excess.

Clean and seal the surface. Wipe off filler residue and dust with a soft cloth. Then seal the wood before the topcoat. The mahogany top on the left was properly cleaned; the right top has been shellacked.

While the squeegee method works quite well at removing the bulk of wet filler from large flat surfaces, turned pieces and intricate moldings are different matters. I've had some luck removing excess filler from turnings using a terry-cloth towel. I've also been marginally successful at removing dried filler from molding nooks and crannies using a shaped scraper. But I often avoid the problem by not filling turned pieces and moldings. The visibility of the pores in these regions is usually disguised because the end-grain wood will finish darker (more absorption) and because of shadows made by the profiles. To prevent filler from getting on these areas, I simply mask them off beforehand (see the top photo on p. 15).

Sand, clean and seal before you finish

Dry time (or more appropriately, cure time) of pore fillers varies significantly. While water-based fillers can usually be sanded and finished within three or four hours, oil-based fillers require two to three days to dry thoroughly. The residual solvents and oils in uncured oil-based filler can cause tiny white spots in the finish if top-coated too soon. This is particularly true when waterborne

finishes and some nitrocellulose lacquers are used.

Once the filler is completely dry, sand down to the sealer, removing all the filler residue from the surface (see the bottom photo on p. 15). Leave filler only in the grain pores. Sand carefully: It's easy to sand through the sealer coat into the base stain. Oversanding can also open up unfilled pores, which will force you to start the whole process over again. Periodically, wipe down the wood with a rag dampened with mineral spirits to inspect your progress. You should wind up with a surface that looks somewhat like the left tabletop in the photo above.

Because grain fillers shrink about 10% as they cure, your freshly filled and sanded wood is probably not going to be silky smooth. You can repeat the process to fill the pores completely, but I prefer to fill the small sink holes with sanding sealer (it's a lot easier). I apply a coat or two of sealer and sand it back to a flush surface. The sealer also provides a good base for the finish (see the tabletop on the right in the photo above). Finally, always make sure your topcoat, sealer and filler are compatible by testing your finishing sequence on scrapwood from your project.

SEALERS: SECRET FOR FINISHING SUCCESS

By Chris A. Minick

Ever try to duplicate the glass-smooth finish that you saw on a fine piece of furniture? Even if you match the stain color exactly, fill the grain pores properly and use an identical topcoat, somehow your finish looks different, or it doesn't feel as smooth. The reason may be that you didn't use a sealer. Understanding why to use sealers and how to apply them will bring a new dimension to your work.

Sealers are the unsung heroes of finishing. For example, high-end furniture often has several layers of finish (usually lacquer or varnish) bonded together with sealers to form a cohesive film. But you would be hard-pressed to know that the sealers are there. When I finished the mahogany tabletop shown in the photo above, I sealed before grain-filling and again before the final finish layer. However, when I started woodworking, I didn't see the usefulness of sealing. It looked like an extra step. Just by dumb luck, the oil-based varnish I used back then worked without a sealer. My early finishes were acceptable, but not great. With time, I began to experiment with different finishing techniques. Several peeling finishes later, I came to realize the error of my non-sealing ways.

Seal first for a better finish. Sealer promotes adhesion and acts as a barrier between separate layers of finish. It can also reduce absorption of the final finish and simplify sanding between layers. Here, Minick brushes a 2-lb. cut of his favorite sealer, super-blond shellac, onto a mahogany tabletop.

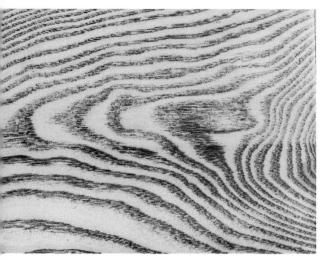

To avoid blotchiness, seal before grain-filling. The author treated the halves of this ash board differently to show the effect of sealing the wood. The dark lower part, which was not sealed before the grain was filled, displays ghost-like smudges. The more even-looking upper part was sealed before the grain was filled.

The best ways to apply common sealers are to brush on shellac, both super blond and orange (left); brush on varnish-based sanding sealer (front); and spray on vinyl-based sealers (right).

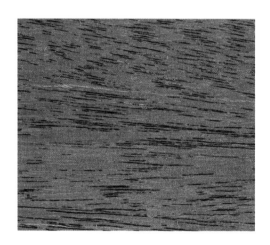

Shellac between finish layers improves finishes. You can sand grain-filler smooth without scratching the under layers, add colored glaze coats without them bleeding, and alternate oil- and water-based layers of finish if shellac is used between layers. Here, shellac sealer helps achieve an instrument-quality finish on mahogany.

Types of sealers

Sealers serve a variety of functions in the woodshop: They ease sanding, decrease finish absorption, promote finish adhesion, and they act as a barrier coat between separate finish layers. Sealers come in several chemical compositions, each tailored to perform a specific task (see the top right photo above). There are three basic sealer types: varnish-based sanding sealers; lacquer-type sealers, including thinned nitrocellulose lacquer and shellac (super blond and orange); and vinyl sealers, which are tougher than the other two.

Sealers make sanding easier

Sanding sealers perform a dual function: They seal the wood and provide a smooth, flat substrate for the final finish. A thin coat of sanding sealer stiffens the wood fibers, so subsequent sanding will cut them off cleanly. The result is a flat, smooth ready-to-finish surface. Most sanding sealers contain metal stearates to make sanding easy. This is the same stuff used on non-loading sandpaper. The soft stearate pigments add volume to the coating. As a result, sanding sealers build fast and dry quickly, but they're relatively soft.

Resist the temptation to use sanding sealer as build coats for your finish; it's never a good idea to apply a thin, hard finish over a thick, soft one. This practice causes increased cold-checking and impact-cracking of hard lacquer finishes. To envision these phenomena, picture a thin layer of ice over soft, unfrozen mud. As you step on the ice, the mud moves, and the ice cracks. Just remember that sanding sealers are meant to be sanded down to the wood before you apply the topcoat finish. If you do this, you shouldn't have problems.

Sealers decrease finish absorption

Finish-thirsty woods like cherry, pine and lauan benefit from a sanding-sealer coat, even if they don't need to be sanded smooth. The stearate solids in combination with the resin in the sealer stuff up the small pores and soft areas in the wood, thus minimizing absorption of the next coat of finish. This is particularly beneficial when you spray on a low-solids lacquer. But, if you use similar reasoning for stain, you can run into trouble. I've seen woodworkers brush sanding sealers on wood before staining in an attempt to eliminate unevenness on blotch-prone woods like pine. I haven't found this helpful. Instead, I use a home-brew of linseed oil as a pre-stain conditioner to reduce blotchiness.

Once you've stained the wood and it's dry, you should seal in the stain layer. This way, you can sand before the next finish layer while the sealer protects the stain from scratches. This is especially helpful if you have to do some grain-filling. Fresh shellac makes a great sealer for this, as does a thinned coat of clear lacquer. But a thin coat of vinyl sealer provides even more protection from sanding abrasion because vinyl sealers are tougher. Sealing before filling the grain will also eliminate smudges that give an undesirable ghosting effect to the wood (see the top left photo on the facing page).

Sealers promote finish adhesion

Oily woods like teak, rosewood and cocobolo contain natural resins that can cause major finishing problems (see the photo on p. 20). Lacquers may peel from the surface or become sticky after they have dried. Worse yet, some oil-based varnishes applied over these woods will refuse to dry at all. Luckily, special vinyl sealers have been developed to make the overlying finish fast, which eliminates these headaches. Vinyl sealers derive their name from the vinyl-toluene-modified alkyd resins with which they are formulated. Vinyl sealers come in a fast-drying lacquer mix for spraying or dissolved in mineral spirits for brushing under an oil-based varnish. Regardless of the carrier solvent, vinyl resins form an impervious layer between the wood and the finish, thus preventing future finish failure. For similar reasons, pigmented primers, such as BIN (William Zinsser & Co.,

173 Belmont Drive, Somerset, NJ 08875; 908-469-8100), are useful when applied under painted finishes.

When you're using vinyl sealer, pay attention to the manufacturer's instructions regarding cure time. Failure to overcoat some vinyl sealers within the specified time can lead to finish delamination. Similarly, vinyl sealers are not really compatible with water-based finishes because water-based resins will not properly adhere to vinyl-alkyd coatings (see the photo on p. 20). But shellac has tremendous barrier properties and adheres phenomenally to both oil-based and water-based finishing materials. Professional furniture refinishers often apply shellac over stripped wood to seal in waxes, silicones and

Incompatible sealer leads to a peeling finish—Always check sealer and finish compatibility first on scrapwood. As the author discovered many years ago on this butternut door, vinyl sealer and water-based polyurethane don't mix.

Sealers increase finish adhesion on oily woods like teak (an unfinished piece is at top). A water-based topcoat knifed with an X shows adhesion differences (from left below): shellac-sealed (good adhesion); not sealed (poor adhesion); vinyl-sealed (poor adhesion). But vinyl sealer is excellent under an oil-based topcoat.

stripper residue that would otherwise interfere with the finish. You can buy shellac premixed, but I prefer to mix shellac fresh using dry flakes and ethyl alcohol. Fresh shellac brushes or sprays on, dries quickly, seals well, is compatible with all common finishes and sands easily. That's why shellac is the sealer of choice in my shop.

Sealing between layers of finish

Sealers allow different finishes to be overlaid on the same project. That's why sealers became an indispensable part of my finishing routine when I started doing multi-layer finishes. For instance, my favorite mahogany finish consists of a yellow ground stain followed by grain filler, three different-colored glaze layers and two or three finish coats. Although I don't use this finish sequence often, when I do, it sure is pretty (see the bottom photo on p. 18).

Here's how the sealer works: Each layer is separated from the next by a coat of shellac. The sealer over the ground stain protects it from abrasion when sanding the filler, and sealer prevents the color from bleeding into subsequent layers. The grain filler is sealed to prevent the porous filler from absorbing color from the first (rosewood) glaze coat. Sealing after this glaze layer keeps it from "walking" into the next (walnut) glaze coat.

Another layer of shellac lets me use an oil-based asphaltum glaze (needed for its color) over the water-based glazes. After I seal the asphaltum layer, I brush on a water-based topcoat. This finish would not be possible without the shellac sealer coats.

A word of caution when you're layering finishes: Make sure all your base coats, topcoats, sealer coats and fillers are chemically compatible. The door in the photo at left is a classic example of what can happen when you ignore this simple rule. I left the peeling water-based topcoat as a reminder of this lesson. Generally, it's wise to choose all your materials from the same finishing family. For instance, varnish sealer and oil-based pore filler can be used under polyurethane. The same philosophy holds true for finishes in the lacquer family and for the water-based finish family. I've had good luck combining oil-based sealers, fillers and stains with water-based topcoats, as long as I seal between each layer with fresh shellac. But the only sure way to tell if your finish layers will be compatible is to test your entire finishing sequence on scrap before you commit it to your project. A little up-front sealer testing can save hours of stripping hassles later.

USING WOOD PUTTY

By Chris A. Minick

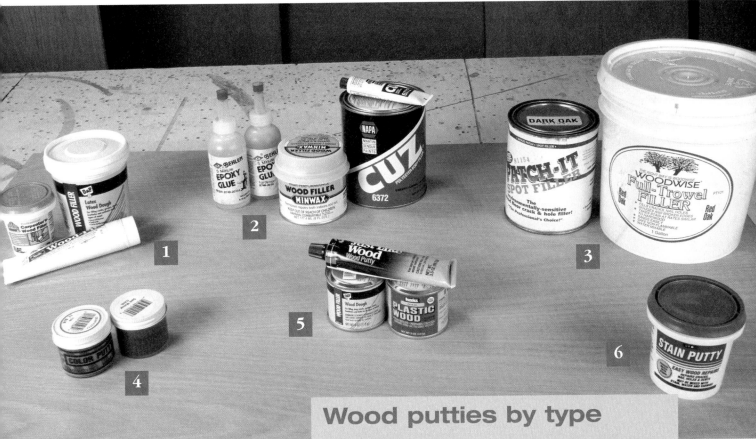

Wood putties by type

1. Water-based putties dry fairly quickly and take stain reasonably well.

2. Two-part mixtures make hard and durable repairs quickly, but they absorb stain very poorly.

3. Floor putties are the author's favorite. They are sold only in large quantities.

4. Oil-based putties are good for minor fill jobs, like finish-nail holes in trim. These putties stay flexible and never dry completely.

5. Lacquer-based putties sometimes shrink and crack as they dry.

6. Dry-powder putty can be mixed with water, stain or clear finishes. You mix only what you need.

In my early woodworking days, a can of wood putty was just as vital as my tablesaw. It seemed as though every project had at least one major putty patch. And regardless of how well I matched the putty color to the wood, a dark blotch always appeared when I stained the piece.

Those experiences taught me two valuable lessons: Store-bought wood putty is never the right color, and the putty always shrinks, even when the label says it won't. My woodworking skills have improved over the years, and obvious defects in my work have decreased to tolerable levels. But that doesn't mean I don't use wood putty anymore. I've just gotten a lot better at hiding the putty splotch.

The kind of putty I use depends on the type of repair I'm making and the kind of finish I intend to apply later. I'll often make my own rather than rely on store-bought versions.

The binder determines the type of putty

Wood putty or wood dough (not to be confused with wood-pore filler) is a thick, pasty material designed to fill nicks, holes or other defects in raw wood. The ultimate wood putty dries quickly, sands easily and takes stain well. Most brands have some of these characteristics, but I've yet to find one that satisfies all the requirements.

Make a putty repair disappear

1. Cover the repair area with sealing tape. To illustrate his technique, the author drills two holes. One will be filled conventionally; the other repair will be camouflaged.

2. Cut and remove a football-shaped patch. With a sharp razor, he cuts an irregular shape around the hole to be repaired. He will fill that and the other hole with latex putty.

3. Sand putty before removing tape. Once the putty has dried, he uses a sanding block to remove the excess.

The human eye has the uncanny ability to pick out regular shapes like circles or straight lines from a random background, while at the same time seeing minor color variations as a single tone. Military camouflage designers have used this to their advantage for years. Break up the lines, add a little color and you can make a tank disappear. I use a trick with putty based on the same principle (see the photos at left and on the facing page). This procedure works best on open-grained woods like walnut and oak. Closed-grained woods like maple and birch take a little more practice.

I start by burnishing a piece of 2-in.-wide clear sealing tape over the defect. The tape prevents the grain pores surrounding the patch from filling with putty. Next I cut around the defect in an irregular shape, cutting through the tape and removing that patch. Where possible, I follow the natural grain lines of the wood. I like to scrape some

I classify wood putties in three broad categories based on their binder resins: lacquer-based, oil-based and latex or water-based putties. Lacquer putties (like Bondex Plastic Wood) and latex putties (like Behlen Wood-Fil) are probably the handiest varieties for the woodworker, but each has its own peculiarities.

If quick drying time is important, lacquer-based putties are the obvious choice. After hardening, they are compatible with most finishes. On the downside, they tend to shrink and crack more than other types of putty, and they take stain poorly.

Latex putties don't have many of the problems associated with putties made from lacquer. Like lacquer-based putties, latex putties dry quickly. But they don't shrink much, have excellent compatibility with finishes, sand easily and accept stain fairly well. They're also easy to tint (see the photo on p. 24). They've become the standard in my

wood from the cutout with a sharp razor knife to give the patch some depth for the putty to fill. For a ¼-in.-dia. hole, I typically cut out a football shape about 2 in. long by ¼ in. at the widest point. I putty the hole with the tape still in place.

When the putty is dry, I sand it down, remove the tape and sand the patch flush. At this point, I apply a base coat of stain to the entire piece because I usually double stain my projects (a ground stain followed by a toning stain). After that has dried, I etch some fake grain marks into the putty patch, trying to mimic the patterns in the wood grain adjacent to the patch. I pay particular attention to spacing and depth. Any sharp scribe will work for cutting grain lines. I've outfitted a mechanical pencil with a sharp needle tip, and I use it exclusively for this purpose. Finally, I apply a second coat of stain that most often hides the patch completely.

4. Remove the tape. The author sands the putty one more time to bring it flush to the surface of the wood after the tape has been peeled off.

5. Scratches give the patch a third dimension. After the first base coat of stain, the author etches grain lines into the surface of the putty repair with a sharp scribe.

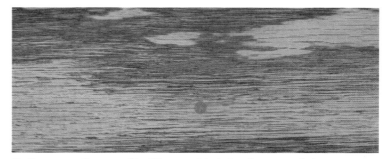

6. Compare the results. The round hole is a lot more obvious at a glance than the irregular patch above it.

Custom colors from store-bought putty—The author used dry pigments and liquid tints to make this color. He dampened the wood with mineral spirits first to approximate a clear finish.

shop. I've found that latex putties sold to wood-flooring installers stain better and shrink less than the ones sold to woodworkers. But these putties are hard to find and usually are sold only in quarts or gallons—more than a lifetime supply for many woodworkers. To find these putties in your area, look in the yellow pages under "Floor Materials and Supplies."

I stopped using oil-based putties in my shop a few years ago after a walnut desk finished with lacquer developed white spots over the puttied areas. Even after three days of drying, the oil interfered with the finish. You can use oil-based putty successfully under an oil finish, but it will remain soft. These nondrying and nonshrinking putties are useful for patching small nicks in finished pieces. I know several custom cabinetmakers who use them for those inevitable installation dings.

Two-part fillers perform some tasks better

Two other putties I've found useful in my shop aren't really wood putties at all—autobody filler (the stuff used to repair rusty cars) and epoxy glue. Chemically, Bondo body filler and the related Minwax High Performance wood filler are two-part styrene polyester fillers. Both can be used to make structural repairs in wood, set quickly and are totally nonshrinking. Best of all, the repair machines like wood. These fillers will not take stain, so they're best used under paint or in a hidden area.

Two-part epoxy adhesive also makes a good shrink- and crack-resistant putty. The slower setting epoxies work better because the extra cure time allows trapped air to rise to the surface. Although most epoxies can be colored with universal tinting colors (UTCs), powdered pigments or fresco colors made for artists work better. UTCs contain solvents that may not be compatible with some adhesives. I have used colored epoxy to

Shopmade putty. The author uses either shellac or hide glue as a binder and mixes it with fine sawdust. The color match is nearly perfect, and the repair virtually disappears.

fill cracks around knots. I've also used epoxy to glue loose knots in place. You can buy UTCs at most paint stores and fresco colors at art-supply outlets.

Shopmade putty for a better finish

Most woodworkers have made wood putty from sawdust and white or yellow polyvinyl acetate glue. Putty made with these glues works pretty well, but it usually doesn't take stain worth a darn. Even unstained, the repair tends to stand out once the finish goes on. To solve that problem, I make sawdust putty in my shop using either shellac or hide glue as the binder.

The procedure couldn't be easier. For projects that will get a clear finish and no stain, I mix fine sanding dust—the finer the better—with shellac until the mixture forms a thick paste, as shown in the photo above. It works just as well as any store-bought putty. Once this is dry, I sand the patch flush and seal the entire project with a

thinned coat of shellac. Because the sealer coat and the binder in the putty are both shellac, the repair virtually disappears. One drawback with using a shellac paste is that it won't take stain evenly.

For projects that will be stained, especially with aniline dyes, I use the same procedure except that I substitute hot hide glue for the shellac. After the defect has been filled, I seal the workpiece with a size made of diluted hide glue, thinned with water to the consistency of milk and applied with a brush. Once that has dried, I sand the surface lightly and then apply the stain. Hide glue absorbs stain amazingly well. The entire piece, including the patch, comes out the same color.

Hiding defects and woodworking mistakes can be frustrating. When all else fails, I ask an artist friend to paint a fake knot over the offending blemish. This method produces the most invisible repair, but it's a bit hard on my ego.

TWO

Coloring Wood

Of the many natural materials man works, few have the intrinsic beauty of wood. Cotton, clay, and iron, for example, really don't look like much until completely transformed into man-made objects such as clothes, china, and hand tools. Wood, on the other hand, can show strikingly beautiful grain as soon as the log is cut open and a surface smoothed. This natural beauty predisposes most woodworkers to leave well enough alone. When they finish a project, they usually reach for something that will let the natural beauty shine through. And in the case of walnut crotch veneer, this is absolutely the best course of action. But it's a rare woodworker who gets to build with walnut crotch veneer every day. For the rest of us, who have piles of pine boards and birch veneer with their relatively lifeless grain, the art of coloring wood beckons.

Coloring wood is an illusionist's art. By means of dyes, stains, and chemicals, humble woods can be transformed into valuable ones and valuable ones made even more striking. It's not a dishonest art, but a frugal one. If you can pay for pine and have mahogany, so to speak, why not? And even more simply, should there be any shame in enhancing wood's beauty? Some say that all wood should be left alone to express its natural beauty; but if so, perhaps we shouldn't paint our cars or houses either and shouldn't wear dyed clothing. Though life would still be beautiful without added color, it wouldn't be as varied or enjoyable. And the illusions of added color and tone aren't only useful to make silk purses out of sows ears. The ability to change the appearance of wood is essential to restoring antiques. The beautiful patina that time and wear give wood can be replicated with the right techniques.

The articles in this chapter will give you a basic understanding of the several ways to alter the color of wood. Dyes and pigments in their various forms are the most common means; but you'll also find articles on ammonia and wood bleaches. Ammonia can substantially darken woods, while bleaches can eliminate unwanted stains (just like in your laundry) or lighten the wood itself. While these articles are by and large overviews of their respective topics, chapter 6 will offer you more specific techniques, recipes if you will, for many of the effects described here.

CHANGING THE COLOR OF WOOD

by Chris A. Minick

Gel stains are great for vertical surfaces like cabinet sides because, unlike liquid stains, they don't run off.

Why would anyone want to stain a piece of furniture or cabinetry and cover up the natural color and figure of beautiful wood? While few of us would even consider staining flame-grained mahogany or burled walnut, not all of us can afford to build every project from first-rate cabinet hardwoods. Most woodworkers I know often employ cheaper woods, such as pine, poplar and birch. And the appearance of such plain woods can be enhanced by staining.

In addition to giving inexpensive woods a richer color, stains are indispensable for matching the color of new woodwork to existing wood furnishings or for evening up natural color variations in boards glued up into wider panels. A judiciously applied coat of stain can even lend a subtle color contrast to bring out the spectacular grain of a highly figured exotic species.

But don't expect to get a perfect staining job by picking up the first can of stain you see on the shelf at your local hardware store and sloshing it on. To get good results for a wide range of staining situations, you need to know the characteristics and qualities of different types of stains, so you can choose the best one to obtain the desired effect. Good stain jobs also depend on proper surface preparation and application technique, so the wood receives the stain evenly. A further assurance of success comes from making stain samples to test the color before applying the stain.

Pigmented stains

Most stains used in modern woodworking shops can be divided into two broad categories according to colorant type: pigmented stains and dye stains. Pigmented stains are

28

suspensions of finely ground colored minerals (mostly iron oxides) mixed into a solvent-based solution. Pigmented stains color the wood when pigment granules lodge in the natural crevices and grain pores on the surface of the wood. This quality makes pigmented stains a good choice for accentuating the grain of ring-porous woods like oak and ash. Unfortunately, the pigment particles will also lodge in sandpaper scratches and boldly reveal a poor sanding job. Pigment particles are opaque; therefore, they resist fading well. They also act like thin paint to obscure the delicate figure of wood like fiddleback maple, making them good for covering up unattractive inexpensive species or plywood.

Most of the stains you'll find on your local hardware store's shelves are pigmented, oil-based stains. The solvent, or vehicle, used in these stains is mineral spirits, and stains also contain a binder (usually linseed oil or an alkyd resin) that acts like a glue to hold the pigment particles on the wood. Without the binder, the dry pigments would simply rub off. The oil binder is the reason you must apply a seal coat, such as shellac, before using a water-based topcoat over an oil-based stain.

Dye stains

Unlike pigmented stains with color particles suspended in a liquid vehicle, dye stains are mixtures of synthetically derived colored powders that are completely dissolved into solution. The color in a dye stain never settles out, so dye stains don't require extensive stirring. Also, unlike pigmented stains, which are made from a limited range of earth tones, dye stains are available in a wide range of hues, including brilliant primary colors. They are ideal for color-matching applications because you can combine exactly the colors you need to make the stain yellower, greener or bluer.

Dye stain solutions penetrate deeply into the wood matrix, coloring each individual cellulose fiber, accentuating the subtle grain patterns in figured woods. However, dye stains won't bring out the contrast in non-figured open-grained woods like butternut and oak, creating a monotone look I don't care for. Dye stains are not as fade-resistant as pigmented stains, so care should be taken

to keep dye-stained wood out of direct sunlight.

Dye stain powders come in three main varieties based on which solvent they're mixed with: water-soluble, oil-soluble and alcohol-soluble dyes. Even though dye stains are often referred to as "aniline" dyes, modern dyes contain no aniline. The name is an unfortunate holdover from 19th-century Germany, where the dyes were first developed from derivatives of aniline (a toxic petroleum-based liquid that's a suspected carcinogen). Rest assured that modern synthetic dye powders are safe to use in the shop.

Stain conditioner prevents a blotchy look

I spent the better part of two months building my first major woodworking project: an Early American-style pine corner cupboard. But when I applied the stain, my would-be masterpiece was instantly transformed into a blotchy mess (even though I carefully followed the directions on the can). I've since learned to eliminate the blotchy stain problem by applying a pre-stain conditioner to the raw wood before applying any solvent-based stain. The stain conditioner evens out the absorbability of the wood, allowing it to take color more uniformly.

Stain controllers made by Minwax and McCloskey are available at most hardware stores, but I home-brew my own conditioner that works fine and costs a lot less. Simply dissolve 1 to 2 cups of boiled linseed oil into 1 gal. of mineral spirits. Brush a heavy coat of the mixture over the entire project, making sure porous areas are kept wet. After 10 minutes or so, wipe off the excess, and follow your normal finishing routine.

Pre-stain conditioners work best on resin rich woods like pine (see the photo at left) cherry or birch. But regardless of species, any parts with lots of exposed end grain (raised panels for instance) will benefit from this treatment, but make a test sample just to be sure.

Pre-conditioning wood prevents blotchy staining. The author's shop-brewed wood conditioner, applied only to the top half of this pine sample before staining, ensures that all areas of the grain will absorb stain evenly.

Water-soluble dyes have the greatest penetrating power of all common wood stains. The deep penetration creates the illusion of depth associated with high-quality furniture. Water-soluble dyes are also relatively resistant to fading, so I prefer them over all others dyes for staining fine hardwoods. And, in case you sand through the finish, water-soluble dyes are more repairable than other wood stains.

Powdered water-soluble dye stains are easily prepared in the shop. Merely dissolve the dye crystals in warm water, let the solution cool to room temperature and it's ready to wipe on the wood. No stinky or hazardous solvents are needed, and cleanup is in warm soapy water. The only real complaint about water soluble dyes is that they raise the wood's grain when applied. But this is easily remedied by wetting the wood before final sanding.

Oil-soluble dyes are closely related to water dyes but are dissolved in a hydrocarbon solvent—usually glycol ether or lacquer thinner. These dyes are often sold pre-mixed as "NGR" (non-grain-raising) stains, so called because the solvent base does not fuzz the grain when applied to raw wood. Oil-soluble dyes form the bridge between pigmented stains and dye stains, giving woodworkers the best of both worlds. But the relatively poor penetration and poor lightfastness of NGR stains somewhat limits their use for fine furniture.

Getting the stain on the wood

There's more to getting a good stain job than just choosing the right color. The final results are determined by how well the wood is prepared (including sanding), choosing the best application method and testing the stain before committing it to your precious workpiece.

Surface preparation
While the degree of surface preparation of raw wood for a clear finish demands fairly standard practices, surface preparation for staining may vary depending on the stain and the wood you choose.

Water-soluble dye stains raise wood's grain and should be applied only after wetting the wood with plain water and sanding the fuzz away after it dries (the grain will not raise again during staining). With pigmented stains, the wood surface must be evenly sanded and free of stray scratches; otherwise, the pigment will show scratches (see the photo on the facing page). This is especially true on close-grained woods that

tend to show scratches anyway. Sand using successive grits, from the roughest to the finest (at least to 180-grit), not skipping more than one grit size between passes. Be especially careful with orbital sanders. Pressing too hard or moving too quickly causes swirl marks that will show up later. Resanding after scratches show up during staining is twice as much work. Certain resinous woods, such as pine, will take stain unevenly even if they've been perfectly sanded, so treat such woods with a pre-stain conditioner, as described in the sidebar on p. 29.

If you're working with dense woods, such as hickory, the degree of sanding affects the amount of pigmented stain the wood accepts, hence the darkness of the final color. It's best not to sand maple with finer than 180-grit paper. Otherwise, the pigment will have no place to stick and the color will look washed out. It is better to switch to a dye stain, which will give you the desired

Stirred vs. unstirred pigmented stains can be vastly different in color. Therefore, always mix thoroughly if you are applying stain right out of the can.

color regardless of how smoothly the surface has been sanded.

Application
There are few restrictions in the way most stains can be applied to wood. You can use a brush, a sponge or a lint-free rag. But avoid paper towels or loosely woven cloth rags that might snag on open-pored woods such as oak. If you own a spray gun and a compressor, spraying can be a time-saving way of applying dye stains to large surfaces, and it's the only way to get an even finish with fast-drying, alcohol-soluble dyes. Avoid spraying pigmented stains. The

Alcohol-soluble dyes are primarily used for tinting or special effects that can be applied with a spray gun. They dry too fast for any other application method. This feature makes alcohol-soluble dyes popular with production furniture finishers. In the small shop, they're normally used only for touch-ups or finish repairs.

Though they are harder to apply evenly than water dyes, alcohol-dye stains have one major advantage over all other stains: They're perfect for tinting or shading wood to create special finishing effects. The best example of this shading technique is the sunburst finish commonly used on guitar bodies.

Any type of dye stain can be a little hard to find locally. The best way to buy them is from woodworking supply catalogs. Woodworker's Supply (1108 N. Glenn Rd., Casper, WY 82601; 800-645-9292) has a finishing supply catalog that has a complete selection of all types of dye stains. If you don't like to order through the mail, try regular fabric dye from your local grocery store; it's basically a dye stain. A pre-mixed, water-soluble dye gel made by Clearwater Color is available from Garrett Wade Co. (161 Avenue of the Americas, New York, NY 10013; 800-221-2942). This product is good for staining vertical surfaces, such as cabinet sides, because the thick gel doesn't run down

Pigment particles lodge in sanding scratches, revealing a poor sanding job. Proper surface preparation requires careful, even sanding, using successive grades of grit, coarser to finer.

abrasive pigment particles can damage the delicate (and expensive) nozzle on your spray gun, literally sandblasting it from within.

Pigmented stains should be thoroughly stirred before application to get the pigment particles that have settled to the bottom of the can back into suspension. Otherwise, you'll end up with a considerably different color (see the photo on the facing page). Because they are true solutions, dye stains can be applied without stirring (I shake them anyhow, just to be sure). Wear gloves and a respirator when applying any solvent-based stain, just as you would for application of a clear wood finish.

The length of time you wait before wiping the excess stain off is relatively unimportant; the final color of the wood is controlled by the concentration of dyes or pigments in the stain formulation. To darken a pigmented stain finish, apply a second coat. To increase the color intensity of a dye stain, increase the concentration of dye powder mixed with the solvent. Incidentally, if you accidentally sand through a dye stain during finishing, apply the original dye solution to the damaged area. The color match will usually be perfect, and the repair will be undetectable.

Test the color first

Evaluating your staining results on raw wood is usually misleading. Stained surfaces look muddy when dry. Dye stains in particular change color considerably once topcoated with a clear finish. And different topcoats will change the final color and sheen of the piece in different ways.

The best way to avoid surprises is to create test panels before staining the workpiece. First, be sure to use the same species of wood as your project—different wood species take stain differently. After applying stain to your test panel, follow your normal finishing procedures, applying all the coats of stain and clear finish, then waxing and buffing your samples. Second, make your samples from larger boards, not small pieces of scrap. I like my test panels to be 4 in. to 5 in. wide and at least 18 in. long. I've found large sample panels to be indispensable for accurately judging the finished appearance of the project.

I save the test panels that look good for reference, with complete finishing instructions written on the back. The ones I don't like are used to heat my shop.

Stains can be mixed or applied in layers. Solvent-compatible stains can be mixed together in the can and applied at once, here over cherry (left sample). Note the difference in the right sample showing the same three stains applied one at a time (from top to bottom): unstained cherry, yellow, reddish mahogany and medium-brown dye stains.

and make a mess (see the photo on p. 28). Bartley's makes a pigmented gel stain, which is available at hardware stores.

Chemical stains

While certain woodfinishers advocate the use of chemicals for changing wood's color, I'm thoroughly against it for several reasons. First, most chemicals used for wood coloring are strong oxidants or are highly caustic and dangerous if they come in contact with your skin. Further, some chemicals, such as potassium dichromate are very poisonous and potentially fatal if ingested. Worse, potassium dichromate looks like a bright orange-colored kiddie drink when mixed in solution.

The second reason to avoid chemical colorants is that they are unpredictable. They create color by reacting with chemicals naturally present in the wood, and results can vary, even in different sections of the same board. Worst of all, these chemicals can deteriorate (oxidize) the clear finish applied over them! Given the low cost and convenience of modern wood stains, there are plenty of reasons to steer clear of chemical colorants.

Mixing different stains together

If you don't see the color you want on the manufacturer's chart, you can often mix stains from different cans in various proportions to achieve the desired color. The catch is that not all stains have the same vehicles (water, oil, alcohol); some types can be mixed and others can't. Further, all oil-based stains will mix with all other oil-based stains regardless of brand. To be safe, you can always restrict yourself to the same brand name and type. The same goes with water-based stains. Manufacturers sometimes mix

two different types of stains together, say, an oil-based pigmented stain and an oil-based dye, for certain colors or special applications. But I'd avoid this practice because it usually brings out the bad qualities of both types and minimizes the good ones.

If you're unsure about the vehicle type of a stain, there are a couple of simple tests you can do. First, smell the stain before it's stirred. It's probably an oil-based pigmented stain if it smells like mineral spirits, and there's a layer of sludge on the bottom of the can (see the top photo at left). In contrast, if a drop of stain in a glass of water dissolves, the mixture is probably a water-soluble dye stain. A drop of oil-soluble dye stain will just sit on the surface of the water.

Layering stains for better effects

If you are trying to match an existing finish of a commercially produced piece, chances are the original stain was applied in layers rather than all at once. Even if you mix exactly the right shade and hue of color in the can, sometimes the results just aren't satisfying on wood. It is not uncommon for commercial finishers to apply a brightly colored dye stain first to bring out the grain, followed by a wood-toned pigmented stain to even up the color. In practice, I often stain the wood initially with a yellow dye stain before applying additional layers of reddish or brownish stains (see the photo at left). I find this tends to bring out the inner figure and heighten the luster of woods, such as cherry, mahogany and walnut.

Another advantage of layering is that it allows you to mix different stains, even if they're not compatible. Nine times out of ten, you'll get away with it. Even if there's a problem, you can try changing the order in which the stains are applied (save any oil-based/self-sealing stains for the last layer). For a dramatic effect, try applying a dark stain to an open-grained wood (such as oak or ash); then lightly sand before applying a lighter-color stain. The dark color remains only in the open grain while the lighter stain colors the surrounding areas, creating a high-contrast effect. Again, experimenting will show you the true effects, and perhaps you'll discover a color effect you couldn't have gotten out of a can.

DRY-BRUSHING WOOD STAINS

by Roland Johnson

I pride myself on being able to restore all types of furniture. So when a customer called on me to look at two grungy, broken-down filing cabinets and asked whether I could bring them up to snuff, I couldn't say no.

The filing cabinets were made of white oak. One was missing a side; the other needed two new sides. The client liked the character of the old pieces but realized they were not valuable antiques. She wanted the repairs done for less than the cost of new cabinets. We discussed options and agreed the new frames would be made of solid white oak, the panels of plywood.

I couldn't get the white oak plywood locally. With the customer's consent, I used

Changing the color of oak—Red oak panels in a white oak frame (left) don't match. So the author stained the piece and dry brushed the red oak to achieve a uniform color (above).

Tinting and toning colors

Color-matching stains can be a real guessing game. A little knowledge about color theory will help make sense of mixing your own stains.

There are three primary colors: red, yellow and blue. Tints are combinations of these primaries. I define tone as the shade (light or dark) of a color. Tint is the actual color.

Let's use red as an example. Red is the tint. By adding black or white, you change the tone. By adding a different color, such as blue or yellow, you change the tint. Pink is a lighter tone of red made by adding white. Purple is a new tint made by combining blue with the red. Equal amounts of all three colors produce brown.

To get specific shades of brown to match wood colors, use more or less of the primary colors. To lighten the tone of your stain, either brush more of it out or thin it with mineral spirits before applying.

Matching dissimilar woods

Every species of wood leans toward certain parts of the color spectrum. In the accompanying article, I matched red oak to white oak. I first blended a stain to match the new white oak to the old, but the stain proved to be too red, or warm, for the red oak panels. To remedy this, I added just a few drops of a blue universal tinting color (UTC) to cool the color and make a good match.

If you have a stain that is a bit on the blue, or cool, side but you want more of a mahogany color, simply add some red tint.

It only takes a tiny amount of colorant in some cases to make large changes in tint. I can usually remedy a bit too much colorant by adding a little bit of the other primary colors to balance my mistake. But the more times I have to add a bit of colorant, the harder it will be to duplicate my efforts.

Keep a variety of tints on hand

My color kit consists of a number of artist's oils for small batches of stains, such as for touch-ups, and less-expensive UTCs for large batches.

The artist's oils I have are burnt sienna, raw sienna, burnt umber, raw umber, yellow ochre, permanent blue, alizarin crimson, white and black.

red oak panels. I now had two finishing challenges: matching new white oak to the aged patina of the original case and making red oak look like aged white oak.

To help make these kinds of repairs appear seamless, I have developed a staining technique I call dry brushing. I've blended the light sapwood of walnut to match the dark brown heartwood. I've used it to even out hard-to-stain woods such as maple and cherry. And I can make new wood look like it's 100 years old.

Dry brushing is a two-step process that begins with traditional staining: The wood is sanded and a stain is applied and then wiped off. When that's dry, a second, heavy coat of stain is applied. This coat is delicately brushed with a soft, dry, natural-bristle brush to remove and blend any excess stain. This method leaves pigment on the surface of the wood as well as in the pores.

A good set of brushes and quality stains and tints

The brush must be pliable and have dense, soft bristles. I prefer natural bristles, but you could use a different kind of brush as long as it's recommended for varnish or enamel. Don't buy cheap brushes; an inexpensive brush may seem soft and supple, but it will be prone to losing bristles. It's not easy to remove bristles from a dry-brushed finish.

I keep a range of brush sizes on hand to suit different jobs. A 2-in. brush works well for small areas such as face frames and chair parts. A 2½-in. brush is good for small panels and other medium-sized surfaces. For large areas, such as tabletops, I use a 4-in. brush. This brush can really move stain around in a hurry.

My favorite stains are oil-based pigment stains produced by Benjamin Moore and Pratt & Lambert. These stains have finely

In UTCs, I keep burnt sienna, raw sienna, burnt umber, raw umber, thalo blue, bulletin red, light yellow, lamp black and white. Using umbers and siennas is a quick way to get basic browns without the need to mix the primary colors together. Umbers and siennas have a tint built in. With a little experience, you will know which to use as a base.

To get a feel for color matching without mixing a batch of stain, practice mixing colorants on a piece of white tag board. I use toothpicks to get a small amount of tint colorant out of the container, and with a small artist's brush, I mix the colors in varying densities to see what changes occur. Make sure you use a new toothpick for each colorant. Just a little contamination can ruin your mix.

Mix and match stains. Benjamin Moore's golden oak and colonial maple stains are mixed to create a tint matching new millwork to an old white oak filing cabinet under repair.

ground pigments and good solvents. Fine pigments help to eliminate brush marks, and good solvents evaporate quickly and evenly. Cheaper stains use solvents that don't seem to have even-drying characteristics. I have not had success with water-based stains because they raise the grain too much.

To create the tints I need, I combine different stains and add tinting colors (see the story above). But you don't need to buy dozens of different stains. I recommend you get a quart each of Benjamin Moore's walnut and golden oak stains. For tinting, purchase 2-oz. bottles of universal tinting colors (UTCs) in red, yellow and blue. These are the basic tints used in paints and are available from most paint dealers. With this kit, you can accomplish a lot.

Because I need to match colors of many different woods in my work, I also use maple stains for their yellow cast, cherry

stains for their red cast and a teak stain for its gray-green cast.

On occasion, a good match using premixed colors eludes me, and I resort to mixing my own stain from scratch. I use a clear stain base (I get mine from a local paint dealer) and color it with artist's oils or UTCs. Artist's oils can be used for tinting small batches of stain, but they are expensive.

Just the topcoat of stain gets dry brushed

To match the white oak frame to the red oak panels on this job, I applied a base-coat stain to the entire piece, wiped it down in the traditional manner and let it dry. Then a second coat of stain, tinted slightly differently, was applied to the panels. These were dry brushed to match the white oak.

I begin by mixing a base-coat stain and testing it on a piece of scrap from the

Apply a base coat. The rebuilt side of the case is covered with a first coat of stain and then wiped off.

Red oak panels get second coat. Apply the blue-tinted stain to the panels; when the stain develops a dull sheen, begin dry brushing. Let the brush just skim the surface.

project. Large differences in grain porosity or wood color—even in the same species of lumber—will affect the results. For the base coat on the filing cabinet, I mixed Benjamin Moore's golden oak and colonial maple stains.

Once I have a good color match, I stain the workpiece (see the photo above). When I stained the new parts of the filing cabinet, I was fortunate that the new white oak millwork blended nicely with the old. But the red oak panels were still too warm.

To adjust a stain's color, I add different tints. To cool down the red oak, I added a little blue tint to the base stain and tested it on a sample. This new batch of stain resulted in a perfect color match between the red oak and the white oak, but the tone was still too light. This is where a dry brushing technique comes to the rescue. I brushed the new color stain over the panels. I let the stain set up until it took on a dull sheen. The time will vary from five to 15 minutes, depending on the temperature.

I brushed the stain back and forth with the grain (see the bottom photo at left), using just the tips of the bristles of a clean, dry, soft brush. The weight of the brush does the work. If you press down too hard (see the top photo on the facing page), the stain tends to move around and the brush gets wet. If you use the sides of the bristles or drag the brush at too flat an angle, the stain will smear and leave obvious brush marks.

It's important to keep the brush dry. I use paper towels to wipe the stain off the tips of the bristles after a few passes. If the brush becomes wet with stain, it will only smear the stain, not dry it. Continue to wipe the stain with the brush until the surface is dry. You know you're done when the workpiece has a uniform sheen and the brush no longer picks up stain. The stain should not show brush marks or any other obvious signs of a thick topcoat. If the results are not to your liking, erase the surface with a rag moistened with mineral spirits.

Overlapping fresh stain over dry-brushed stain can be a problem. The fresh stain's solvent will dissolve the built-up pigment of the dry-brushed stain quickly, resulting in a poor blend line. Always try to find natural breaks to stop and start the brushing, and try to work small areas at a time. The only exception is a tabletop. Here I do the entire surface at once. I work fast, but I never hurry. On a piece that is fairly complex, such as a chair, I tend to do one or two parts at a time. Sometimes I'll mask off completed areas to avoid getting fresh stain on an already brushed surface.

Spray on a protective finish

A dry-brushed surface needs a protective coating. Any solvent-based finish will work, but you must apply it by spraying. A dry-brushed surface is very delicate because pigment is floating on top of the wood. If you try to brush on a finish coat, solvents will dissolve some of the dry-brushing, and you'll have a real mess. Handle the piece carefully before final finishing.

I spray my work with an acrylic lacquer. I start with one coat of sanding sealer, lightly sand with 220-grit and then apply two coats of finish, sanding between them with 220- or 320-grit. If you don't have spray equipment, you can use aerosol cans of spray sealer and finish.

Too much pressure—This will only sweep the stain around.

Just the right touch—Gently sweep bristles across workpiece.

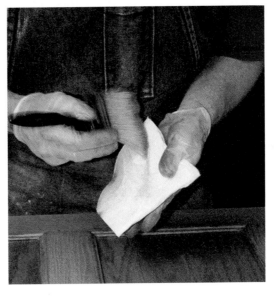

Keep the brush clean. Wipe bristles every few strokes.

MIX YOUR OWN OIL STAINS

by Tom Wisshack

Stain can bring out the best. The author's table, veneered with crotch mahogany and built with cherry legs, received just one light coat of his homemade oil stain. After observing the natural colors already in the wood, he mixed a stain that accentuated them and gave the wood a head start on developing a patina.

I'll be the first to admit it. There's a real purity to a "natural," unstained wood finish, a real virtue to letting the wood's true figure and color come through. But if you are refinishing, restoring or reproducing a piece of furniture, well, a "natural" finish is something that you just can't afford. Color, tone and patina take years, sometimes decades, to develop on "naturally" finished pieces. In almost 20 years of refinishing and restoration work, I have developed a way to get the right color and patina in a matter of hours.

My technique for coloring wood is better than either aniline dyes or commercial stains because of the control I have over tone and depth of color. Also, the stains are largely reversible. I make my own oil stains with turpentine or paint thinner, linseed oil, Japan drier and artist's oil colors.

The turpentine serves as a solvent, diluting the pigments in the artist's oil colors; the linseed oil acts as a binder to keep the ingredients in solution; and the Japan drier ensures that the oil colors will dry within a reasonable amount of time (some dry much slower than others).

One exception is that I substitute copal painting medium (available in art-supply stores) for the linseed oil if I'm working on an antique. The reason is that linseed oil will tend to darken most woods over time. The copal works just as well as a binder. When working with an antique, I take another precautionary step. I also seal the surface prior to staining with shellac before applying any stain, so the stain can be removed entirely at a later date if more work is to be done on the piece.

The key to my stains—the secret ingredients—are the artist's oil colors. What makes them so special are the quality of the pigments used and the fineness of the grind. Artist's oil colors are generally ground much finer than the pigments used in commercial stains, which are often the same as those used in paints. Because the pigment particles are so fine, the resulting stains are much more transparent than commercial stains, letting more of the wood's figure and grain show through. And artist's oil colors are permanent and more fade-resistant than off-the-shelf wood stains.

An infinite range of color choices is one good reason to make your own oil stains. A sample board illustrates the subtle colors possible using artist's oils for your pigments:

A. The first section is natural Honduras mahogany with just one coat of linseed oil.

B. Section B has a light coat of the author's homemade oil stain applied to it. The stain consists of turpentine, linseed oil, Japan drier and just a bit of burnt-umber oil color.

C. More umber has been added to the same stain to produce the tone in section C.

D. Cadmium red and yellow are added to the same stain to heighten the colors already in the wood.

E. Finding section D somewhat too red, the author added a little green to neutralize the red and to bring the tone back to brown.

F. A little black adds depth to the stain.

G. The mixture was thinned with turpentine to yield the natural-looking result in section G.

Getting the color right—Mix artist's oil colors separately on a sheet of glass, and then add them to a mixture of turpentine or paint thinner, linseed oil and Japan drier until you have the tone of the pigments you want. Copal painting medium should be substituted for the linseed oil whenever you don't want to darken the wood, such as when refinishing an antique.

Mixing the stain

I mix the liquid ingredients in a glass jar. For a small batch of stain, I'll start with about a pint of turpentine or thinner, one-third cup linseed oil and three or four drops of the Japan drier. I mix the artist's oils separately on a small sheet of glass (my palette), and then I add the mixed pigments to the liquid mixture a little bit at a time until I get the depth of color I'm looking for. I adjust the mix of pigments to get the tone I'm after (see the photo above).

I'm looking for a very dilute stain, on the order of a tenth or so as concentrated as a commercial product but with the consistency of low-fat (2%) milk. The advantage of such a dilute stain is that I can control it by applying it in two or three coats rather than all at once, deepening the tone while still retaining a semitransparent surface. Additionally, if the color is not quite right, I can adjust it repeatedly to alter the tone without ending up with a muddy, murky mess.

The maximum amount of artist's oils I add to the 1-pint solution is about one-third of a standard-size tube, or a little less than half an ounce. This can vary, depending on the intensity of the colors used, so you'll have to experiment. But even the finest quality artist's oils will give you an opaque finish if you get too heavy-handed with them. More light coats are better than fewer heavy coats.

Because these stains are so dilute, it's rarely necessary to seal new wood prior to staining. An exception is pine, which may appear blotchy regardless of how dilute the stain is. A penetrating sealer, such as one of the commercially available Danish oil finishes or a thinned solution of tung oil, eliminates this problem.

Applying the stain

I generally brush on the first coat of my homemade stain, let it stand about 20 minutes and then wipe it off. Leaving the stain on the wood for more or less than 20 minutes will not dramatically alter the amount of color the wood absorbs but how you wipe off the stain will. A brisk rub leaves only traces of the stain on the wood's surface. Gently wiping in circles and then with the grain will leave considerably more stain on the wood. Subsequent coats can be applied with a cloth.

If you don't like the way the stain looks on the wood, usually you can remove most of it with steel wool and naphtha or paint thinner while the stain's still wet. After the wood has dried, you can try again.

Quick custom oil stains from Japan

by Mario Rodriguez

When building an antique reproduction or recreating a missing component, an important and difficult part of the job often can be the precise matching of the original's color. It's almost impossible to achieve this with the application of a single coat of stain even if you mix your own stains. The task often requires several coats, with successive coats used to deepen or adjust the previous application of color. My system of alternating a light coat of lacquer between coats of stain gives me unparalleled speed, flexibility and reversibility.

For my stains, I use Japan colors suspended in turpentine. Japan colors are highly concentrated basic pigments, usually in an oil-based solution. They're used widely as tinting agents and are available in a variety of colors from most woodworking catalogs with a good selection of finishing products. A ½-pint can generally costs from $7 to $12. Each can will last a long time because the pigments are so concentrated and because I'm only using small amounts of any given color, always mixing several colors together to get the precise tint I need for a job.

I can custom mix practically any shade I need by combining two or more colors and can control the intensity and opacity of the stain by varying the proportion of Japan colors to turpentine. I have used this technique to alter harsh or unnatural colors from commercial stains. Vivid or garish reds and oranges, for example, can be dramatically changed to cooler browns and rusts with a light wash of green. Or you can go the other way: I've also warmed up plenty of dull gray-brown walnut pieces with a light red-orange wash.

I mix my stains by pouring a little more turpentine than I'll need for the job into a glass jar and then adding the Japan colors to the turpentine. I check the color and intensity of the stain on a sample board and then adjust accordingly. I apply the color most often with a rag, because this eliminates any lap marks. But I use a brush when I have to get the stain into tight, otherwise inaccessible areas. I leave it to dry thoroughly (usually no more than a couple of hours, but I always check that it's dry in cracks or anywhere the stain could pool).

After the stain is completely dry, I spray on a light coat of lacquer to act as a sealer or barrier coat. To apply the lacquer, you can use a conventional spray rig, an HVLP (high-volume, low-pressure) unit or even aerosol spray cans if you don't have spray equipment. I've had good luck with Behlen's Jet Spray Furniture Finish (available for $5.95 a can from Garrett-Wade (161 Avenue of the Americas, New York, NY 10013; 800-221-2942).

When the lacquer dries (it takes no more than twenty minutes), another coat of stain can be applied to darken or change the color without disturbing the previous layer of stain. If the second coat of stain doesn't achieve the color or effect you want, simply wipe it off and try again. With this technique, achieving perfect color matches or exact colors is just a matter of patience and experimentation.

Sealing in the stain

After staining, I like to allow at least three or four days (a week is even better) before applying a finish. This allows the stain to dry thoroughly, minimizing the chance of it bleeding into the finish. An additional precaution I often take is to use a dilute coat of dewaxed (the most refined version, also called blond dewaxed) white shellac as a sealer between the stain and whatever I decide to use for a finish. The shellac will isolate the oil stain so that practically any finish can be applied without problems. Or you can just use the shellac itself as the finish.

Sometimes I'll also "spice" the white shellac with orange shellac. I add it in small increments to give the surface an amber tone that's reminiscent of an older piece. Whatever finish you use, though, be sure to refer to the can or the manufacturer's instructions to make sure it's compatible with the shellac sealer.

FOR VIBRANT COLOR, USE WOOD DYES

by Chris A. Minick

Getting consistent colors—Exact measurements and careful record keeping are important for duplicating colors. Wood dyes usually are a blend of colors, visible as dye powder dissolves in water (right) and in filter paper (above).

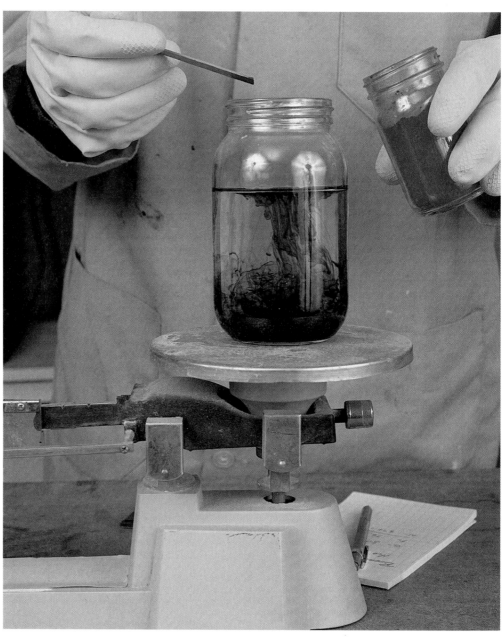

Aniline dyes are a good product with a bad name. Their nasty reputation is a holdover from the days when these versatile coloring agents were highly toxic. It's a misnomer not much different from golfers still calling their drivers "woods," even though many modern golf clubs are made of metal. Woodworkers still know dyes by the name "aniline," even though modern wood dyes no longer contain the chemical.

First used in the textile industry in the mid-1800s as a substitute for natural dyes, aniline-derived dyes worked fine, but they faded quickly and were soon replaced by more light-fast synthetic colorants. Unfortunately, the term aniline dye stuck. It is still used to distinguish transparent wood stains from their pigmented cousins.

Dyes are useful for special finishing effects, like layering (adding depth) and toning (applying tinted finish). Probably the best use for dyes is evening out differences in color, like those between sapwood and heartwood.

Dyes can work miracles on figured wood (see the photos on p. 44), but they aren't magic. For example, when an uninteresting piece of wood is dyed, it will just become an uninteresting, colored piece of wood.

You can buy premixed-liquid or gel wood dyes or mix-it-yourself powdered dyes. I mostly use powdered dyes, which have an indefinite shelf life. Dye is classified by the solvent that dissolves it. The three classes are water-soluble, oil-soluble and alcohol-soluble dyes (see Sources of Supply on p. 49). Each type has finishing advantages.

You can even use ordinary fabric dyes. Brands like Rit can be found at department stores, but you'll have to mix or layer several colors to get more natural wood tones. Powdered fabric dyes sometimes have fillers, so I buy the premixed-liquid type. They're fairly inexpensive, so they're good for experimenting.

Dyes are less hazardous than many household cleaners, but you will still need to handle dyes carefully:

- Use a paper mask when mixing the dye.
- Wear rubber gloves, so you don't absorb the dye through your skin.
- Keep dye powders and solutions away from children and pets.
- When a dye is mixed with a flammable solvent, store it properly.
- If you get dye on your clothes, wash them separately.

Differences between pigments and dyes

What distinguishes dye stains from pigment stains is the size of the particle that's doing the coloring. Individual colorant particles in a dye solution are exceedingly small—there are more than 10 million trillion per quart. In comparison, the particles in pigment stains would look like boulders.

Pigments are suspended when in solution; dyes dissolve totally in solvent. The tiny size of dye particles explains why dye stains are so transparent and why they penetrate wood so deeply. Pigments stay near the surface of wood where they lodge in wood pores, which emphasizes the pores and any blemishes like sanding scratches. Dyes color everything similarly. Even end grain can be dyed so that it looks like the rest of the wood.

Water-soluble dyes have lasting color and clarity

Water-soluble dyes are the most versatile of the three wood-finishing dyes. Water-soluble dyes are easier to use, easier to repair and are more light-fast than the other two types. The exceptional clarity and penetration of water-soluble dyes help make figure come alive. Laboratory experiments confirm that water-soluble dyes penetrate the wood about five times deeper than alcohol-soluble dyes. The deep penetration and chemical structure of water-soluble dyes account for their superior fade resistance. (The story on p. 45 gives a general explanation of how fading occurs.)

To mix water-soluble dyes, I use a gram scale to weigh the water and dye powder (see the right photo on the facing page). Keep track of dye brands, colors and concentrations every time you use them. If you ever have to match a color, a mixing logbook will save you hours of making up sample

Quartersawn lacewood—Burnt-sienna and then light-walnut dye bring out the ray-fleck figure.

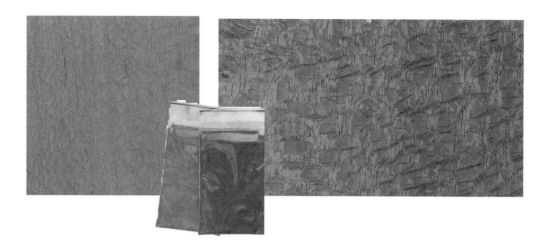

Mottled mahogany—Yellow and then rosewood dye emphasize the chatoyance.

Quilted bigleaf maple—Scarlet-red and then cherry-brown dye highlight the undulating figure.

stain boards. Once the dye is mixed, sponge the wood with the solution until the wood is thoroughly wet. Wipe off the excess before it dries. (Leaving wet dye stain on wood for a long time will not darken the color any further.)

Because water raises wood grain and makes the surface fuzzy, water-soluble dyes do the same. Fortunately, there is a simple solution to this. I flood the wood with clear water after I have sanded to 180-grit. After the wood dries overnight, I knock down the raised grain with 220-grit sandpaper. Once the grain has been sanded flat, the dye stain will not raise the grain again.

Dyes go deep but still fade

Pigments tend to obscure wood's fine details. By contrast, dyes are more transparent, which lets the wood show through. Instead of muddying subtleties in figure, dyes enhance them, as shown in the photo at right.

Even though dyes penetrate more than pigments, dyes fade more. Fading is a form of photochemical degradation. Though ultraviolet light plays a part in fading, intense visible light is mainly responsible.

Visible light is composed of seven colors: red, orange, yellow, green, blue, indigo and violet. White light is a blend of all these colors. A red dye stain looks red because the dye absorbs the other colors and reflects only the red.

Dyes are large, organic molecules primarily composed of atoms of carbon, hydrogen, nitrogen and oxygen. The arrangement of these atoms within each molecule dictates how a dye responds to light.

Quite often, enough light energy is absorbed by a dye molecule to initiate a photochemical reaction, which changes the arrangement of its atoms. Photochemically changed molecules usually are colorless. Because of this, the color becomes more dilute; therefore, the dyed wood appears lighter—faded. Pigments produce color the same way as dyes, but they are more immune to fading.

Alcohol-soluble dyes fade the fastest. The alcohol-dyed half of the sample shown in the photo at right faded from a nice walnut color to swamp-green in less than two months under fluorescent lighting. Water-soluble dyes fade the least (see the unfaded portion of the photo at right). Oil-soluble dyes fall somewhere in between.

Even though the fading of dyes is inevitable, don't let it prevent you from using them. If you use a fade-resistant dye, your project should remain the same color for decades.

Pigment (top) vs. dye stain

Faded (top) vs. unfaded

Blotch-prone woods like cherry and pine don't fair any better with water-soluble dyes than they do with solvent-based pigment stains. To minimize blotchiness, I substitute a hide-glue size for the initial coat of clear water. Make the glue size fairly dilute (by weight, I use one-part hide glue granules to nine-parts water). If you use premixed hide glue, you'll have to dilute it as well. Once dry, the size accepts the dye stain evenly. This only works with hide glue, though. I once ruined a butternut desk by trying white-glue size.

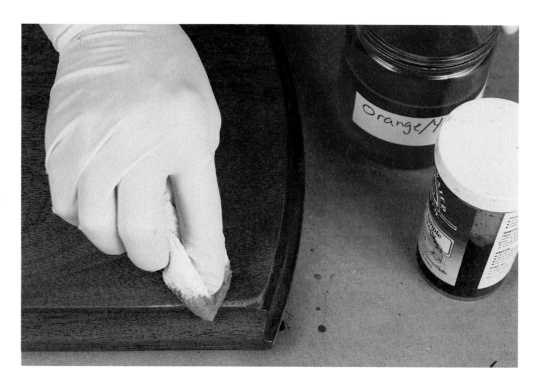

Use dyes for finish touch-ups. The author stains a sand through on a mahogany tabletop. After he applies an orange-red dye, he'll seal the repair with shellac. Once that's dry, he'll wipe on the rosewood dye and seal it in preparation for a topcoat.

Oil-soluble dyes can customize stain color

Most woodworkers have used gallons of oil-soluble dye over the years and don't even know it. Pigment-stain manufacturers often include oil-soluble aniline dyes in their stain formulations to add a little life to an otherwise dull stain. Oil-soluble dyes will dissolve in common shop solvents, like mineral spirits or VM&P naphtha, but they dissolve most completely in lacquer thinner.

Once dissolved in solvent, oil-soluble dye can be added to linseed oil, Danish oil or varnish to make a custom color. In solution, oil-soluble dyes can also be added to a can of pigment stain to modify the color. One problem with oil-soluble dyes is their lack of clarity. Because of their muddy look, I don't like to use oil-soluble dyes on raw wood. But I still keep a full array of colors in my shop for tinting varnishes when I'm toning areas of furniture.

Alcohol-soluble dyes tint shellac, lacquer

Comedian George Burns once asked a clothing-store clerk what the shrink-resistant label on socks meant. She replied, "The socks will shrink, but they really don't want to." The latest alcohol-soluble dyes, touted as "fade-resistant," are some-what analogous to this.

I've found that most of these alcohol-soluble dyes will fade, but they really don't want to. They do have a place, though. Furniture restorers like them for tinting shellacs and solvent-based lacquers in touch-up work.

Alcohol-soluble dyes can be dissolved in methanol (wood alcohol) or ethanol (grain alcohol). I like ethanol because it's the least toxic of the two. Alcohol-soluble dyes dry very rapidly, so they can leave lap marks when brushed or wiped on. Spraying is really the only acceptable way to apply them to large surfaces. Because most alcohol-soluble dyes fade quickly, I find little use for them in my shop.

Non-grain-raising dye stains save sanding

Dye stains that do not raise wood grain are called NGR (non-grain raising) stains. Although NGR stains are technically not a separate class of dye stains, many woodworkers view them as such. But here's the rub: Some brands (the bad ones) are just oil-soluble dyes dissolved in solvent. They give wood the bland look of oil-soluble dyes. Good brands of NGR stains, like Behlen's Solar-Lux (see Sources of Supply on p. 49), are water-soluble or organic dyes that, through a chemical sleight of hand, offer decent clarity and penetration without mak-

ing the wood fuzzy. If you drop some NGR stain in clear water and it dissolves, it's a good one.

NGR stains made with water-soluble dye still lack the depth of penetration of water-dissolved dyes, so they look a little flat by comparison. There are rare occasions, though, when a water-soluble dye is impractical. Intricately carved areas, for example, can't be sanded easily after a water-soluble dye has raised the grain. For these situations, I'll use an NGR stain. I make my own by mixing concentrated powdered dye with hot water and then diluting the solution with lacquer retarder from James B. Day & Co. (1 Day Lane, Carpentersville, IL 60110; 708-428-2651). A volume ratio of one-part dye solution to three-parts retarder is about right.

Adjusting dye color to suit the wood

Customizing the color of a dye stain is easy. All dyes within a solvent class can be intermixed. For instance, any two water-soluble dyes can be mixed or layered to produce a third color (see the photo on the facing page). Likewise, colors within the alcohol-soluble and oil-soluble families of dyes can be blended.

Dye colors are not always consistent from one supplier to the next or even from one batch to another from the same company. Luckily, you can modify the color slightly by adding small amounts of liquid dye-tinting colors.

I use Dayco brand (carried by James B. Day & Co. and most professional paint-supply stores).

You can also tint dye to get that special color you want. A dull-looking walnut can be livened up, for example, by adding a bit of red tint. Adding green to a cherry dye stain (which is often too red) will cool the overall color to a more natural cherry tone. Conversely, dyes that are too blue can be warmed by adding orange dye.

Color intensity (how light or dark a dye stain is) is controlled by the amount of solvent in the dye solution.

So if your dye stain is too light, just add more dye powder. I add a little black India ink to my dye stains when the standard

Dyes are great for special color effects. After building a case for his son's electric guitar, the author custom-finishes the lid. Successive bands of color create a sunburst effect.

color is a little too bright and needs to be toned down a shade or two. India ink is not a dye, but rather a dispersion of very fine lamp-black pigment that imparts a neutral gray tone to dye solutions. Incidentally, quartersawn walnut stained with India ink makes a decent substitute for ebony.

Special effects: layering, shading and toning

Woods with large, open pores like oak look a little strange when stained with dye. The areas between the grain lines color evenly, but the open pores do not. Dyed oak usually lacks contrast between the earlywood and latewood bands.

I solve this problem by layering a pigment stain over a dye stain. I start with a yellowish-brown, water-soluble dye, seal it with shellac (let it dry) and then wipe on walnut-colored pigment stain. The shellac prevents the walnut stain from coloring the areas between the grain lines. But the pigment does color the open pores. The result looks like antique oak.

The basic idea behind layering is to create distinct depths of color within the wood. Layering different dye stains produces an effect that can-not be achieved any other way. Dye-layered finishes look particularly stunning on wooden instruments.

One of my favorite layered finishes is for mahogany. I start by applying a bright yel-

low dye stain to all surfaces. This first layer, called a ground stain, highlights the figure deep in the wood and evens the color of the separate boards that make up a piece. The next layer is a coat of rosewood dye stain made by Clearwater Color Co. (Highland Hardware, 1045 N. Highland Ave. N.E., Atlanta, GA 30306; 800-241-6748). The rosewood dye gives the wood a rich, red-dish-brown hue. The topcoat of finish can even be tinted to bring out other highlights. The timing of the dye applications is critical to getting distinct layers. For instance, I apply the second dye when the ground-stain looks dry but feels damp. The second dye does not penetrate as deeply as the first, so two layers of color are formed.

As I mentioned, certain dyes are soluble in finishes. Oil-soluble dyes can tint oil finishes and oil-based varnishes. Alcohol-soluble dyes can tint shellac and lacquer. Water-soluble dyes and NGR stains can tint waterborne finishes. Because all these dye-tinted finishes are transparent, two fancy techniques, toning and shading, are possible.

Toning is applying a tinted finish to an entire piece to alter the overall color slightly. Shading is more of a decorative effect that's achieved by selectively applying a tinted finish to highlight areas of a piece. Shading the center of a tabletop darker than the edges, for example, gives the table a worn, aged look.

You can improve your dyeing methods with different applicators. With small brushes, for example, you can color in areas of wood or add detail, as shown in the photo on p. 47. With a spray gun, you can cover large areas or add zones of color (see the photo above).

But the best advice for using dyes, no matter how you apply them, is to experiment with a dye stain on scrap until you're happy with the color.

If you absolutely hate the results, don't despair. You can sponge on full-strength chlorine bleach, and the color will disappear.

SOURCES OF SUPPLY	Water-soluble dyes	Alcohol-soluble dyes	Oil-soluble dyes	NGR stain
H. Behlen & Bros. 4715 State Highway 30 Amsterdam, NY 12010; (518) 843-1380	✓	✓		✓
Furniture Care Supplies 5505 Peachtree Rd. Chamblee, GA 30341; (800) 451-0678	✓	✓	✓	✓
Garrett Wade Co. 161 Ave. of the Americas New York, NY 10013; (800) 221-2942	✓	✓		✓
Homestead Finishing Products 11929 Abbey Rd. N. Royalton, OH 44133-2677; (216) 582-8929	✓	✓		
Lee Valley Tools 1080 Morrison Dr., Ottawa Ont., Canada, K2H-8K7; (800) 461-5053 (U.S.)	✓			✓
Olde Mill Cabinet Shoppe 1660 Camp Betty Washington Rd. York, PA 17402; (717) 755-8884	✓	✓	✓	✓
Woodcraft 210 Wood County Industrial Park Parkersburg, WV 26102; (800) 225-1153	✓			
The Woodworkers' Store 4365 Willow Dr. Medina, MN 55340; (800) 279-4441	✓	✓		
Woodworker's Supply 1108 N. Glenn Rd. Casper, WY 82601; (800) 645-9292	✓	✓	✓	✓

GLAZES AND TONERS

by David E. Colglazier

Glazing transforms color and adds detail. Glazes are colored finish layers applied over a sealed base, like this painted cabriole leg. Glazes stay workable long enough for blending and texturing.

Many woodworkers assume they're committed to store-bought stain colors. For some finishing jobs, though, a one-time application of stain just won't do. But by adding colored finish layers at the right time, you can alter or compensate for an existing color as you go, getting exactly the right result. Two finishing products, glazes and toners, will let you do this.

Glazing and toning can add depth and color to a finish or adjust the hue to get the look you're after. I rely on both methods in my antique-restoration work because there's no other finishing process I'm aware of that

Glazes make a leg look old. The author used glazes on three legs of a table to match a leg that had darkened from iron reacting with tannin. He applied a tan base color and then defined the pores and grain patterns with darker glazes.

Wiping off a glaze changes the look. To show how color and texture can dramatically change, the author brushes and then wipes off a burnt umber glaze on one of the oak legs. Mineral spirits or naphtha can be used to soften or remove an oil-based glaze layer. A sampling of brushes used for texturing is in the background.

can bring such subtle refinement or dimension to a finish. Despite their similarities, glazes and toners are used differently.

Glazes rely on an applicator to add texture or simulate grain detail. It helps to think of glazing as painting (see the photo on the facing page) because you're covering, or at least partially obscuring, a base color of some kind. Glazes usually go on just before the topcoats so that you won't disturb or cover up the brushstrokes.

Toners are generally not manipulated with a brush or rag after they are applied. Think of toning as applying thin layers to alter the overall color of a piece. Spraying is best.

Glazes and toners are great for refinishing, restoration and color matching, but they aren't for every job. They require more artistic skill than other finishing methods. With glazing and toning, you need to know how to spray a finish. You often have to lock in a layer of glaze or toner by spraying a coat of nitrocellulose lacquer or shellac. If your shop isn't equipped to do this, you can use aerosol cans of lacquer (made by Deft) and shellac (Wm. Zinsser & Co.), which are readily available.

Layering is the key

The human eye is a very perceptive tool. With training, it can observe at least five variables of a finish: surface defects, wood-pore and flat-grain color, finish depth, top-

coat sheen and texture. Glazes and toners rely on the eye's ability to perceive depth. By visualizing what the final result will look like two or three steps ahead, I can plan glaze and toner layers that will compensate for or correct a hue that isn't quite right. (The story on p. 55 gives a brief explanation of color matching.) Each layer, whether opaque, transparent or somewhere in between, affects the final color, texture and readability of the underlying wood.

Layering a finish is like building a house from the foundation up. Layers can be applied in many orders, but some are more practical than others. From the wood up, this might be a finish-layering sequence: tint and apply pore filler, dye or stain to get the right flat-grain color, correct the hue with a toner or semitransparent glaze, lock that in with a clear layer, add a thicker glaze for texture, tone where needed to add color or shade, and put on the topcoats. Toners can be added just about anytime in the layering process to change the overall color because, usually, they are nothing more than tinted finish. However, if you want to apply a heavy, textured glaze, you typically would apply it at the end of the layering. Unlike toners that can serve as their own barrier layer, glazes always need to be topcoated.

What are glazes and toners?

Glazes and toners are special stains meant to be applied over a sealed surface, rather than applied to bare wood. Glazing stains come as liquids in cans and are most often brushed or wiped on with a cloth. Toning stains come in aerosol cans (see Sources of Supply on p. 55). The pigments used as colorants in glazes make them opaque. Toners usually are a lacquer-based solution of dye and/or pigments. They're almost always thinner and more transparent than glazes, but here's where the terminology can get confusing. What some finishers call toner, others refer to as shading stain. Likewise, glazing is sometimes called antiquing. To distinguish some of the terms, I put togeth-er a glossary of common colorants (see the box below).

Great for restoration jobs and color matching

Old finishes are not uniform. They become worn in places, faded in others. They accumulate dings, dirt and wax from being used and polished over the years. To match the finish of an old piece of furniture, you have to fake the patina it has acquired, which can be complex. Mixing up trial stains could get you the right color, but stains ordinarily are used directly on the wood. Once applied, they are difficult to remove. By contrast, glazes and toners are layered over a sealed base (see the drawing on the facing page). Glazes can add an unusual color or mimic a grain pattern. Toners can blend in a repair, hide a wood defect or create a special effect, such as shading. I use toners more than glazes, though I often use a combination of both in the same project.

Glazes and toners could be useful if you want to make new work look old or add a special look to a new piece, like a sunburst. Glazes and toners conceivably could give more mileage to an undesirable piece of wood. For example, a glaze could be applied to a board to simulate figure. Or, to get wider stock for a panel, you could tone the sapwood so it matches the heartwood.

Glossary of common colorants

The two most common colorants are pigments and dyes (not including substances that chemically alter wood color, such as bleaches). Pigment and dye stains can be applied to wood or as colored layers of finish.

The definitions (to simplify things, I omitted paints) at right are partially adapted from several manufacturers' literature and from Bob Flexner's book *Understanding Wood Finishing: How to Select and Apply the Right Finish,* Rodale Press, 1994.

Pigments: Ground opaque particles that, when added to a binder, color wood at the surface, lodging in pores, scratches and defects. Pigment stains vary from semiopaque to semitransparent and fade slowly. Pigments are a key ingredient in glazes.

Dyes: Tiny particles that color wood or dissolve in finish to add a transparent color layer. Dyes penetrate deeply but are known to fade. Because of their clarity, dyes offer good depth and grain readability. Dyes are often used for toning.

To lock in a layer, use a barrier

Glazes and toners can be layered one over the other or separated by a clear film (barrier) of finish. When you don't want to disturb what's underneath, you should spray on a barrier layer. I use nitrocellulose lacquer mostly and sometimes shellac. I avoid waterborne lacquers because they can cause compatibility problems.

A barrier can lock in a layer of color and let you, with care, alter a subsequent layer without damaging what's under it. Lacquer barriers or lacquer-based toners can help melt one layer into the next. If a glaze layer doesn't look right, it can be removed with a rag dampened with the appropriate solvent (mineral spirits or naphtha for an oil-based glaze). Each glaze, toner and barrier layer should be thoroughly dry before you do the next. Be especially careful when spraying lacquer over oil-based glazes because wrinkling can occur if each isn't allowed to dry thoroughly. I use several thin coats of lacquer or shellac, so any solvent will evaporate completely. Certain shellacs can introduce yellowing; however, that might be what I need to give the piece a golden, aged look.

Glazes are applied and then manipulated

Glazes develop a bite on an undercoat as the solvent evaporates, but they still offer plenty of working time (5 to 10 minutes). I apply

Anatomy of a layered finish

A layered finish can add depth to a piece, adjust color, obscure or pronounce detail, add an aged look and permit easier repair to the finish. The order of the layers can vary. The illustration shows just one sample.

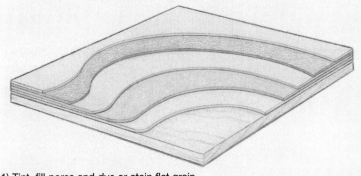

1) Tint, fill pores and dye or stain flat grain.
2) Apply sealer (may be colored).
3) Add toning layers and barrier layers (if needed).
4) Use glazing layer for final color adjustment and surface texture.
5) Apply topcoat(s) to seal, give protection and add sheen.

the glaze over the surface and work it until the brush starts dragging (see the photo at left on p. 51). This happens as the glaze turns flat. I can use a brush or rag to remove glaze from the high spots, leaving it in the recesses (see the photo at right on p. 51).

Sometimes I use a dry-brushing technique, which is glazing with an almost empty brush. The bristles stay soft, not tacky or stiff as they would if the glaze were drying. Dry brushing offers the most control

---

Stains: A broad label applied to any mixture of pigments, dyes, resins and solvents that alters wood color. The percentage of pigment affects the clarity: Glazing and pickling stains are semiopaque, pigmented stains are semitransparent and penetrating stains are quite transparent.

Glazes: A fairly thick oil-, varnish- or water-based stain that contains pigments. Glazes are usually brushed or wiped over a sealed surface and spread or partially removed as (or just after) the thinner evaporates. Glazes are used for antiquing, coloring pores, accenting grain patterns and adding depth to carvings and turnings.

Toners: Fast-drying solution (usually lacquer) containing dyes and/or pigments applied to a sealed surface to alter the color. Toners are sprayed on the entire surface and left to dry. Pigmented toners tend to obscure the under-color and detail; dye toners are more transparent.

Shading stains: Designed for highlighting, shading stains are specialized toners that are applied to specific areas. They can give a shaded appearance to a surface or blend regions of color. Tinting lacquers are similar products that build quickly and are used to unify tones.

A glaze patina— The author applies dark glaze to a corner block for an old door frame to emphasize its age. After a light wash coat, he can dab on heavier coats in the recesses of the rosette and nail holes to simulate an accumulation of dirt.

Toning unifies an antique sofa table. The author often tones and glazes furniture parts separately. Here, he sprays the legs and stripped table edge with a red mahogany toner. He used pigment from a can of dark stain to glaze the edges of the stretcher. The legs were wiped with this glaze, left to dry and then shellacked.

Toner used as a shading stain— To simulate a table with a faded center, the author shades the edge of this mahogany top with a dark toner. After he rings the top with light, even coats, he can refine the look and color by spraying other toner bands.

for putting down a minimal amount of glaze. To soften an oil-based glaze, I apply mineral spirits or naphtha after the bite occurs. This gives me a bit more time to experiment and is especially useful when I'm matching wood patterns or texture.

Viscous glazes applied over a nonporous surface can be manipulated with rags or brushes to produce special effects. Marbleizing, graining, faux-finishing and antiquing are all forms of glazing. Glazing brushes come in an assortment of sizes and bristle types. Many finish-supply stores carry a good selection of them.

I prefer oil-based glazes because of better compatibility between brands and because the solvents don't rapidly affect the previous layers I've applied. To get started, it's a good idea to practice with just a couple of glazes from one product line. Then you can expand your range with confidence. As you get better, you can use glazes in more creative ways (see the top photo above).

Toners are sprayed on and left to dry

Toners come in many pigment and dye combinations ranging from opaque to transparent. Transparent toners can be layered to adjust color without losing the distinction between the pores and the flat grain. I probably use transparent toners the most. They're ideal for shading (see the bottom photo at left) and for blending colors on components of an original piece (see the center photo above). Using opaque toners can be like glazing. The color becomes muddier and the wood lacks grain definition, but this can be an advantage when, for example, I need to disguise a blemish. The thickness of the layer can be varied to get more opaqueness, too.

You can make your own toners by mixing dry pigments and/or alcohol-soluble aniline dyes in shellac or lacquer. For toning (shading) specific areas of furniture, I mix up a shading stain using lacquer and a low concentration of dye. I apply the shading stain in three or four thin layers so that I can sneak up on the color and not overdo it. I can always add another light layer, but if the color is too dark, it's nearly impossible to lighten uniformly. Every job hones your application skills and perception of color.

SOURCES OF SUPPLY

Constantine
2050 Eastchester Rd.
Bronx, NY 10461
(800) 223-8087

Liberon/Star Supply
P.O. Box 86
Mendocino, CA 95460
(707) 937-0375

Mohawk Finishing Products, Inc.
(H. Behlen & Bros.)
Route 30 N.
Amsterdam, NY 12010
(800) 545-0047

Olde Mill Cabinet Shoppe
1660 Camp Betty Washington Rd.
York, PA 17402
(717) 755-8884

Star Finishing Products, Inc.
360 Shore Dr.
Hinsdale, IL 60521
(708) 654-8650

The Woodworkers' Store
4365 Willow Dr.
Medina, MN 55340
(800) 279-4441

Woodworker's Supply, Inc.
1108 N. Glenn Rd.
Casper, WY 82601
(800) 645-9292

Color matching made easier

I often have to match colors that a client or a decorator has selected. It can be tricky finishing a piece so it goes well with a rug, the wallpaper, the couch fabric, the curtains and the other wood in the room. There are three things that make my job easier: a color wheel, stain-sample sticks and the proper lighting.

Color correction is the art of knowing which color additives are needed to make a certain hue. For instance, red can warm up brown, and green can cool it. As simple as this sounds, the permutations of hue become far more numerous by adding black and white to darken or lighten the color.

Interestingly, men have more difficulty at color matching than women because more men have color blindness in the red and green regions of the spectrum. I don't have this problem, but even so, I still need help with color decisions. I use a primary color wheel. Grumbacher

Stain sticks aid color choices— Guided by a fan of stain sticks, the author chose a glazing stain for this tabletop. The samples also helped the customer come up with a color that makes the veneer band look natural and blend with the chair fabric.

wheels (called Color Computers) are available from Star Finishing Products (see Sources of Supply at left). The wheels come with directions and a summary of color theory.

Stain sticks, a collection of stir sticks that are already stained, are also helpful. The sticks (I use Old Masters brand, but you can make your own) are pinned at one end like a set of feeler gauges. I can fan them out (see the photo above) and ask the customer to determine the color direction. I don't have to make up a wall full of sample boards.

Back at the shop, I try to match colors under the same light that will be used to view the piece. True colors can change as a result of the light source. For example, incandescent light is rich in red; fluorescent light is predominantly blue. A balance of cool-white and full-spectrum fluorescent bulbs is pretty close to sunlight.

Recently, I replaced the fixtures in my shop with T8 lamps made by Philips, which use triple-phosphorous tubes. The tubes are very efficient. The light has a warm color temperature and a more natural look in the shop. They've made color matching much easier.

FUMING WITH AMMONIA

by Kevin Rodel

Anyone who's spent time mucking out stables, or just walking through a working barn, knows how pungent ammonia fumes are. Those fumes have darkened the beams of many a barn over the centuries. I wouldn't doubt that many farmers put two and two together when they noticed how quickly oak acquired an aged patina.

Around the turn of the century, fuming became popular with many of the furniture-makers and manufacturers working in the Arts-and-Crafts style. So much so that when most people think of Stickley, Limbert or Roycroft furniture, fumed white oak is what they see in their mind's eye. Other woods can be fumed, but white oak responds best and most predictably to fuming (see the bottom photos on p. 59). For a look at the effects of fuming on other woods, see the box on p. 60.

Regardless of species, boards that will be fumed should all come from one tree. Different trees within a species will vary in their tannin content because of growing conditions. This will affect how they react to the ammonia. Because it's difficult to get boards all from one tree at a regular lumberyard, I buy most of my lumber from specialty dealers who saw their own.

I began fuming furniture because I'd become increasingly interested in the Arts-and-Crafts movement. I had been making more furniture in that tradition, and I wanted it to convey the look and feel of the originals. The finish seemed like an important element in the whole equation. Fuming is not the perfect colorant for every situation and wood species, but where it does work, it works very well and can give a superior finish to stains or dyes.

Stains obscure the surface of the wood somewhat. Worse yet, on ring-porous woods like oak, pigments collect in large open pores, making the rings very dark and overly pronounced. The effect is quite unnatural and looks to me like thousands of dark specks sprinkled across the surface. Also, stains are time-consuming to apply, and I have a strong aversion to exposing myself to the volatile fumes of the petroleum-based products found in most commercial stains.

Aniline dyes do a better job than stains, but they're also rather labor-intensive and can be very tricky to apply well. Dyes also fade over time, especially in direct sunlight. Fumed wood is colorfast.

The thing I like best about fuming is that what you see after the process is still only the wood, just as clearly as before. It's just darker. That's because the ammonia reacts with tannins that are naturally present in the wood, actually changing the color of the wood, not merely adding a superficial layer of color. Samples of fumed wood that I've cut open show a ragged line of darker wood between $\frac{1}{16}$ in. and $\frac{1}{8}$ in. deep.

Another thing I like about fuming is that it's virtually foolproof. The first piece you fume will look great. Unlike stains or dyes, fuming won't make a piece look blotchy or cause drips. And there's one other benefit to fuming. While the piece of furniture is being fumed, you can get back to work. The ammonia keeps working while you're taking care of other business.

Handle ammonia with care

The first and most important consideration when fuming is safety. Before you even buy the ammonia, make sure you have a properly fitted face-mask respirator with ammonia-

Fuming with ammonia gives white oak that classic golden-brown color. Before it's been fumed (above), white oak is a pale, almost cool, tan.

Aqueous ammonia is poured into a glass container placed at the bottom of the fuming chamber (above). Then the top of the chamber is lowered quickly onto its base (right). Protective gear is essential.

filtering cartridges. Other types of cartridges, such as those used for spraying lacquer or other finishes, are not designed to filter ammonia fumes and will not offer protection. Ammonia cartridges are inexpensive and available at any fire or safety equipment store. Look in the yellow pages for the one nearest you.

Eye protection is essential. I use swimming goggles, which fit tightly around the eyes. The purpose of the goggles is to protect the eyes from fumes, not just accidental splashes. Rubber or plastic gloves are also necessary. Read the precautions on the side of the ammonia bottle, too.

Finally, if you're trying this for the first time and you work in a basement shop, wait until the weather is nice and do the fuming outside. After you become comfortable with

the procedure, you can consider doing it indoors.

The reason for all the precautions when fuming is that ammonia used for fuming wood is not common household ammonia. It is a strong aqueous solution that has between 26% and 30% ammonium hydroxide. Household ammonia has less than 5%.

You'll want to buy the ammonia locally and pick it up yourself. Because it is considered a hazardous substance, shipping charges are high (more than the cost of the ammonia). This industrial-strength ammonia is used in machines that reproduce blueprints and surveys, so you can usually find it at business-supply, blueprint-supply or surveyor-supply stores (look in the Yellow Pages for a supplier). It's sold by the gallon.

Here in Maine, it costs between $6 and $10. And 1 gal. fumes a lot of furniture.

Bringing ammonia and wood together

With safety equipment and ammonia in hand, you're almost ready to fume. All you need now is some kind of fuming chamber—the more airtight the better. The most versatile and efficient chamber construction seems to be a heavy-gauge (3 mil or greater) plastic wrap stapled to a simple softwood frame that's held together with drywall screws (see the photo at right on the facing page).

This type of chamber is lightweight, can be made to just about any size and can be broken down into flat panels for storage. If a fairly large chamber is needed, one side panel can be used as a detachable doorway. Use spring clamps or hand screws to attach the door panel and felt weather stripping as a gasket to seal the chamber. Small chambers can be placed over the items being fumed, as in the photo at right above. If you're fuming outdoors, be sure to weight or tie down this kind of chamber. They're very light and blow over easily.

I've used many other types of fuming chambers as well—everything from large plastic trash cans (perfect for small items) to a rented moving van. The van allowed me to fume an entire bedroom set at one time for a reasonable cost. The ammonia did no harm to the van, and by the time I returned it the morning after fuming, there was little if any residual smell. And because every piece was exposed to the ammonia for the same amount of time, I was able to achieve a precise color match.

Prepare a piece of furniture to be fumed the same way you would for staining or finishing. Scrape or sand until the surface is smooth, and remove any hardware. Place the piece of furniture in the chamber so that no part that will be visible is touching anything. If the ammonia vapors can't circulate, they won't be able to react with the tannins in the wood. As a result, that spot will not darken like the rest of the piece.

Never let the furniture come into direct contact with the aqueous ammonia because it is very corrosive. I use glass pie plates to hold the ammonia. They're relatively inexpensive, clean up completely and can be used over and over again. They also present a large surface area to the air so the ammonia evaporates readily.

I fill a plate about half full and place it on the floor of the chamber (see the photo at left on the facing page). The plate should be filled quickly but carefully. If you're fuming a particularly large piece or more than one piece, you may want to use two or three pie plates. Attach the door to the chamber, or lower the chamber onto its base. With the fumes confined to the chamber, you can remove your mask and goggles. Note the time so you can keep track of the exposure.

Test pieces determine color

The length of time a given piece will need to be fumed depends on the volume of the chamber, the amount of ammonia used, the species of wood being fumed and the depth of color you're looking for. Knowing when to remove a piece is largely a matter of personal experience. You can hedge your bets, though.

White oak (unfumed to 32 hours exposure)

Fuming common furniture woods

The practice of fuming wood to enhance its color is most often associated with white oak. The oaks in general are high in tannin and fume well, though red oak tends to turn greenish rather than deep brown like white oak. Other species contain varying amounts of tannins and can be fumed, but the effects are generally not as pronounced as with white oak. I was curious about the effects of fuming on other furniture woods, so I fumed a number of them for four hours.

I'd heard that nontannic woods could be fumed if a solution of tannic acid was applied to the surface of the wood first, so I tried that as well. (Tannic acid is available from Olde Mill Cabinet Shoppe; 717-755-8884.) Tannic acid is sold as a powder that you add to water. I added tannic acid to a pint of water until the solution was saturated, applied the solution with a foam brush and then let the samples dry overnight before fuming. Here are the results.

The best way to know when you have achieved the desired amount of fuming is to use test pieces. I always place three or four pieces of scrap, preferably cutoffs from the same project, on the floor of the chamber. When I think enough time has gone by, I don mask and goggles, quickly open the chamber, remove one of the scrap pieces and reseal the chamber.

When it first comes out of the chamber, the wood will have a gray, almost weathered, look. Don't be alarmed; this is normal. To see an approximation of what the finished piece will look like, I apply a coat of finish. As soon as the finish goes on, the real color imparted by the fuming appears instantly, almost magically. If I want the piece darker, I'll continue checking the color of the scrap boards at regular intervals until I'm happy with the result.

If, after eight hours in the chamber, a piece is still lighter than you'd like, you should replace the ammonia. I put on my mask, goggles and gloves, open the chamber and dump the old ammonia into a bucket of water. I add fresh ammonia to the pie plate, reseal the chamber and leave the bucket of diluted ammonia outside for a day. Then I pour it around the trees in our orchard or on the compost heap.

Once you've decided the wood is dark enough, remove it from the chamber, and let the piece of furniture off-gas for 8 to 12 hours. I try to plan my fuming sessions so that the piece comes out of the chamber at the end of the work day. By morning, there's little residual smell.

At this point, you can apply your finish. Oil, varnish, shellac—any finish will work. There's no problem with compatibility between a piece of furniture that's been fumed and the topcoat. At the same time, fuming doesn't protect the surface of a piece in any way, so build up your finish as you would normally.

My preferred finish has always been boiled linseed oil (I use Tried and True brand because it builds quickly and contains no metal driers). Three or four coats over fumed oak impart a subtle amber overtone that's in keeping with the look of Arts-and-Crafts furniture.

	Unfumed	Fumed	Tannic acid, fumed
Maple — No finish / Oil			
Birch — No finish / Oil			
Cherry — No finish / Oil			
Butternut — No finish / Oil			

USING WOOD BLEACH

by Jeff Jewitt

M y client's dining table had been damaged in a move, and two of its leaves were missing. The French-style reproduction table, about 60 years old, was veneered with a fruitwood that looked like cherry in grain and texture. But the wood had mellowed to a yellow-gold color. Cherry was the natural choice for the new leaves. But the color would be too red and would darken significantly over time.

I solved the problem with bleach. It removed the natural color of the cherry, providing me with a neutral background so I could match the original with a dye stain. The bleach also halted the darkening process in the cherry leaves, so the color of the table would remain uniform.

Matching old wood to new is only one application for wood bleaches. Most finishers are aware that bleaches remove unwanted stains—food, black water and old dyes. But bleaches can do much more. They also even out tonal variations in dissimilar woods and produce blond or pickled finishes. The trick is knowing which bleach to use. For that, it helps to understand how wood bleach works (see the box on p. 64).

For woodworkers, there are three general types of bleaches: peroxide, chlorine and oxalic acid. All three work by altering the way wood molecules reflect light, thereby changing the color in the process. But each type of bleach is suited to particular tasks; they are not interchangeable.

Ideally, a bleach should work selectively to remove color, meaning that it should only remove the color that you want and not the color of anything around it. In most cases you'll need to experiment, especially if you don't know the composition of the stain.

Pick the right bleach for the job. These three types of wood bleach all have specific uses. No one bleach does it all.

Because most bleaches are highly poisonous and often very corrosive to skin, you should always wear good rubber gloves, a dust mask (if you're mixing dry bleach powders) and safety glasses.

Peroxide bleaches remove natural color

These bleaches are sold as two-part solutions, commonly labeled A and B. You'll find peroxide wood bleaches in most paint

and hardware stores. The two chemicals are usually sodium hydroxide and a strong hydrogen peroxide solution. When used together, a powerful oxidizing reaction takes place that is effective in removing the natural color in wood, like the mahogany shown at right. To a lesser degree, peroxide bleaches will lighten some woods that have been treated with pigment stains. They are ineffective on dye stains.

The most common way to apply this product is to wet the wood thoroughly with sodium hydroxide (part A) and immediately follow with hydrogen peroxide (part B). It's important with some tannin-rich woods like cherry and oak that part A not sit too long before part B is applied because the sodium hydroxide may darken the wood. You can also mix the two parts together and apply them at the same time, as long as you do this quickly after the parts are mixed. Usually one application is needed, but a second application may be necessary to even out the bleaching effect.

Some dark woods, like ebony, are not affected by peroxide bleaches. You can use this to your advantage if you want to bleach a tabletop with ebony inlay. On some woods, especially walnut, a greenish tinge may appear in some areas if the bleach is applied unevenly. To prevent this problem, apply the bleach sparingly; use just enough to make the wood wet. Don't flood the surface.

Neutralize the alkaline effect of peroxide bleaches after the wood has dried by applying a weak acid, like white vinegar. Use one part vinegar to two parts water. Follow that with a clean water rinse.

Peroxide bleaches will remove all the natural color variations in wood, so use them judiciously. I use them to match sun-faded wood or to provide a neutral base for a decorative finish like pickled oak. You can also use them to compensate for heartwood/sapwood variations, but I usually prefer to bring the sapwood in line with the heartwood by hand coloring or spraying the sapwood with a dye stain.

Peroxide bleaches work best at removing the natural color of the wood. This piece of mahogany veneer changed from a deep red to a light blond color with one application of peroxide bleach.

Chlorine bleaches eliminate dye colors

Chlorine is a strong oxidizer that will remove or lighten most dye stains (see the bottom right photo below). A weak chlorine-based laundry bleach such as Clorox will work, but it will often take several applications to be effective. A much stronger solution can be made from swimming pool bleach—a dry chemical called calcium hypochlorite. It's inexpensive and can be purchased from a retailer of pool supplies.

The chief advantage of chlorine is that it will remove or lighten the dye without affecting the natural color of the wood. You can use laundry bleach or the stronger version—dry calcium hypochlorite powder mixed to a saturated solution in hot water. A saturated solution is created by adding the powder to water until no more powder will dissolve. Mix only in glass or plastic containers: The chemical will attack aluminum or steel. The mixture will lose its effectiveness if stored, so I make up only what I'll use right away. Cool to room temperature before using, and filter out solids.

Apply the solution liberally to the wood and, in some cases, the dye will immediately disappear. Some dyes may take longer to bleach, and some may only lighten but not disappear. Wait overnight to determine the full bleaching effect. If the color hasn't changed after two applications, applying more bleach won't help. You'll need to try another technique. Chlorine bleaches are usually ineffective on pigment-based stains.

Chlorine bleaches remove dye stains. The natural color of walnut (left) is virtually unchanged by an application of chlorine bleach. But most of the dye stain on the birch veneer (right) has been removed with the same solution.

How bleach works

Color in an object is produced when the molecules selectively reflect light. These colored molecules may be organic, like those in dyes, or they can be inorganic, like those in pigments. Most bleaches, like peroxide and chlorine, work by disrupting the way that the molecules can reflect light. Other bleaches, like oxalic acid, convert the colored compound of a stain to a different, colorless one. The physics of these concepts may be difficult to understand, but the important thing to remember is that bleaches do not really remove the color of a substance. They simply change the material so it appears colorless.

As an example, tannic acid and ferrous sulfate when dissolved in water are colorless solutions. When mixed together, the two chemicals react and form a third compound, iron tannate, which is a grayish-black color. Iron tannate is the compound responsible for most of the black water spots on oak. When oxalic acid is added to this liquid, it converts the colored iron tannate molecules to iron oxalate, a colorless compound. When used in this respect, oxalic acid is a bleach.

Not every colored object can be bleached. Colors that are produced by inorganic molecules will not react to the bleach. Many pigments like carbon black (used in inks) and earth pigments (used in wood stains) will not react to bleach. These colors can only be completely removed by scraping or sanding the color off the surface of the wood.

The only way to remove these are by sanding or scraping.

Oxalic acid for iron stains and weathering

Oxalic acid is unique in that it will remove a specific type of stain formed when iron and moisture come into contact with tannic acid. Some woods, like oak, cherry and mahogany, naturally contain a high amount of tannic acid. A black stain results when the tannic acid reacts with water containing trace amounts of iron. Oxalic acid will remove this discoloration without affecting the natural color of the wood (see the top photo on the facing page).

Oxalic acid also lightens the graying effects of outdoor exposure. It is the active ingredient in some deck brighteners. If used on furniture that has been stripped for refinishing, it will lighten the color and re-establish an even tone to the wood.

Iron-based stains are fairly easy to spot. They are grayish-black and usually ring-shaped. They may also show up as a splotchy appearance on oak that has been stripped. Before applying oxalic acid, remove any finish first.

In a plastic container, mix a solution from dry crystals of oxalic acid (available from most woodworking supply stores) in hot water. Allow the solution to cool to room temperature, and apply it to the entire surface, not just to the stain. Several applications may be needed with overnight drying in between. Once the surface of the wood is dry, any residual oxalic acid must be removed before sanding or finishing because the acid will damage subsequent finishes. Several water rinses will remove most of the oxalic acid crystals left on the wood surface. Neutralize the acidic wood surface with a solution made from one quart of water and two heaping tablespoons of baking soda. Then rinse off the baking soda solution with water.

Solving special staining problems

Stains that form on wood during the drying process are varied in their composition. Sticker stain, brown stain, streaking and light "ghost" stains are all common problems. Some can be removed by bleach. The composition of the stain may be chemical or biological, so a trial-and-error approach may be needed when attempting to remove a stain. I often start with oxalic acid and then follow with chlorine. Peroxide bleaches are a last resort because the removal or acceptable lightening of the stain can result in bleaching the surrounding wood.

Stains like grape juice, tea and fruits can be removed with a chlorine bleach. Remember to wipe the entire surface to get an even effect. Some blue and black inks with an iron base can be eliminated with oxalic acid, but carbon-based inks, like India ink, can't be removed by any bleach.

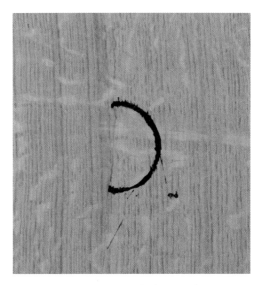

Oxalic acid removes the stain but not the color of the wood. The author used a deck brightener, which has oxalic acid as an active ingredient, on this piece of white oak.

THREE

Hand Finishing

It's often said that there is no better teacher than experience, that book-learned or second-hand knowledge pales in comparison. Every beginning woodworker knows that it's a lot easier to read a description of how to cut a mortise by hand than it is to do it. All you have to do is hold the chisel square and hit it with a mallet dead on, but it's never really that easy the first time around. The same is true of finishing techniques. Padding a lacquer finish is really just a matter of rubbing the little pad back and forth over the board, right? If you've tried it, you know better. Somehow the pad always gets stuck to the board, leaving a crater of gummy finish. Each wipe removes as much finish as it puts down.

The solution is obviously to get some experience. Once you've applied finishes a few times, their mysteries fade into the background while simple details and techniques take center stage. However, just going at the finishes a bunch of times isn't very helpful because you'll keep making the same mistakes. To guide you through those first few attempts, this chapter presents the accumulated experience of many professional and amateur finishers who have spent years working the kinks out of their work, and then took the time to write about what they do. Some of the finishes they present would indeed be difficult, if not impossible, to figure out without their directions. But with them, even a high-gloss varnish finish becomes a simple matter of following the steps and paying attention to the details. No mysteries. Few difficulties.

This chapter collects a range of the most common and accessible finishes that can be applied by brush, rag, or pad. Most are presented in a step-by-step fashion that makes them devilishly easy to replicate in your own shop. Mario Rodriguez's careful description of padding on lacquer is one such. He makes a near cousin of the French polish very approachable, if almost easy by breaking down the process into small steps. The first time you try this finish, just keep the book open on your bench and read through as you work, as you would if you were cooking from a recipe. Other articles show great techniques for getting the most out of penetrating oils, the bird's eye low-down on shellac, and a neat way to bring out a shine with wax. And if you don't get one just right the first time, don't fret. With the experience of the first try, the second can't help but be better.

FINISHING BRUSHES

by Jeff Jewitt

Applying finish with a brush seems easy enough. Dip the brush into the finish, spread the finish on the wood and then wait for it to dry. That's the theory, anyway, but many woodworkers are disappointed with the brush marks, streaks and bubbles that can mar a finish. Maybe, they may wonder, there's some secret technique. Or maybe the finish itself is to blame. Quite often, though, the problem is neither the technique nor the finish. It starts with the selection and use of the brush. Using the wrong brush or a second-rate brush makes it difficult to get first-rate results.

A brush is more than some bristles attached to a handle. Brush-making is an art. Manufacturers mix bristles of different lengths and stiffnesses for different types of brushes. In a top-quality brush, the bristles are selected and arranged by hand. For a closer look at the parts of a brush, see the photos and drawings on the facing page.

Manufacturers of cheap brushes economize on the content and configuration of the bristles. They may use an oversized divider to give the brush an illusion of fullness (see the photo on the facing page). Bristle tips on a good-quality brush have natural splits, or flags, that help hold and spread the finish. Brushes that are cut to shape after they are formed are cheaper to make, but they will be missing flags at the bristle tips. That's a good indication the brush won't perform very well.

The most important, and the most expensive, brush component is the bristle. The type of bristle determines the suitability of a brush for a particular finish as well as how it works in general. Bristles can be divided into two broad categories: natural animal-hair and synthetic-filament bristles.

Animal hair is best for solvent-based finishes

Natural-hair brushes are expensive and don't perform well with water-based finishes. But for top-quality results using oil-based varnish or paint, natural hair is unsurpassed.

Natural hair is divided into two categories: stiff bristle and soft fur. Hog bristle is used in most painting and finishing brushes. Soft fur, such as sable, camel, ox, skunk or badger, is used for varnish and artist's brushes. Two or more types of hair are often combined for specific performance characteristics.

Hog bristle is for paint and varnish.

Chinese hog bristle (also called China bristle) is the best. The natural split ends on these stiff bristles allow the brush to carry a good deal more finish than bristles with smooth tips. The natural taper toward the tip gives hog bristle its strength and resiliency, or spring, which is especially important when applying paints and varnishes. The paint or varnish can be worked into the pores of the wood with the tip of the brush.

Sable is for detail work.

Sable is the best natural hair for artist's brushes. Sable forms a fine, strong point when wet, making it ideal for touch-ups. Kolinsky sable is the best and most expensive; hairs from other red weasels are cheaper. All are known as red sable.

Reservoir

The divider is a wooden cleat that creates a reservoir and gives shape to the brush. Artist's brushes don't have dividers.

The setting is an adhesive, usually epoxy, that holds the ends of the bristles in place.

The ferrule is a metal band that joins the setting and the handle.

The handle usually is a dense hardwood, such as beech. Handles are shaped for different applications. Long, thin handles are for precise control, and short beavertailed handles are for better balance and less fatigue.

Brush on the right is better—These two rectangular, chisel-edge brushes are made from similar parts, but the one on the right is of better quality. A thick wooden divider at the center of the brush on the left gives it an appearance of fullness, but a heavier divider means fewer bristles. The brush on the right has a smaller divider and more bristles, so it will hold and release finish more evenly.

Flagged tip

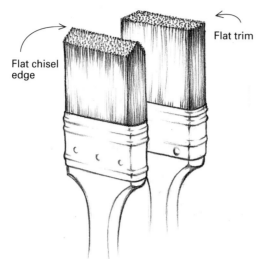

Flat chisel edge

Flat trim

Bristles are the most important part. Look for flagged tips, which help hold and spread the finish.

Camel is soft

A camel brush is good for lettering. The fur is not really from camels but usually from the tails of Russian and Siberian squirrels. Other kinds of squirrel hair are too coarse. Cheaper grades of camel brushes are made with ox, goat or pony hair.

Badger is best for oil-based finishes

This very soft and resilient hair is regarded as the best for flowing on oil-based finishes. It does not have the body of hog bristle, so it's usually combined with a coarse hair, like skunk or black bristle. Pure badger-hair brushes are used for blending and highlighting in glazing and wood graining.

Ox is for lettering

Ox hair is taken from behind the ears of oxen and is silky and durable. It resembles sable but cannot form as fine a tip. It's used in lettering and sign painting.

Fitch is a combination of hair and bristle

Fitch is a confusing term because it applies to both a hair and a type of brush. American fitch is skunk hair. European fitch comes from a gray or black weasel. Fitch brushes usually are a combination of hairs—skunk on the outside for softness and bristle on the inside for stiffness. Fitch brushes are

excellent for flowing finishes, such as oil-based varnish.

Synthetic bristles best for water-based finish

Synthetic bristle is a good choice for all types of finishes. It hasn't eclipsed natural bristle for the ultimate varnish brush, but synthetic bristle is constantly improving. The search for synthetic filaments to replace natural hair has been ongoing since the beginning of this century.

The first synthetic filaments were blunt tipped, similar to toothbrushes. Nowadays, manufacturers use several filaments that are tapered like natural bristle. Du Pont's Tynex and Chinex are manufactured specifically for brush-making. Taklon (a generic name) is a dyed white nylon filament with a tapered shaft and a smooth, unflagged tip. It is used extensively in artist's brushes.

Chinex is the most recent synthetic filament and is good for oil-based finishes and excellent for water-based finishes (see the bottom photo on p. 73). Taklon artist's brushes are exceptional for applying all finishes and usually are available in sizes up to 1½ in.

The chief advantage of synthetic filament over natural hair is that synthetic filaments absorb only 7% of their weight in water. Hog bristle and natural hair may absorb as

The author's choice

Here's a selection of brushes that I've found to be most useful in my finishing business.

Taklon synthetic bristle brush has a tapered filament without natural splits at the ends of the bristles. This brush is made without a divider, so it has a very narrow chisel edge. I use it on small projects where exceptional control is needed. It can be used with all types of finishes. I recommend a ¾ in., 1 in. or 1½ in. brush.

Red sable is great for touch-up work. I recommend at least two sizes, a #1 and a #4.

Combination badger/skunk brush is first rate for all flowing finishes, particularly oil-based varnish and polyurethane. A 2-in. brush will cover most tasks. An expensive brush, but it's well worth it.

Chinex synthetic bristle is a good all-purpose brush. It's my favorite for water-based finishes. A 2-in. brush will cover most applications. It's an excellent tool for applying water-based dyes and stains and also can be used for solvent lacquer and oil-based varnishes.

China bristle brush is less expensive than fitch and excellent for all oil-based finishes. I use 1 in. and 1½ in. for detail work and 2½ in. and 3 in. for large surfaces. I particularly like oval-shaped brushes because they hold a lot of finish. Large oval brushes are called varnish brushes; smaller ones, 1 in. or less, are called oval sash. They also work well for effects like glazing, dry brushing and highlighting.

Cleaning a brush ————————————————

Start by wiping off the excess finish on newspaper. Then dip the brush into the appropriate cleanup solvent, and squeeze out the excess. Pour a liberal amount of dish-washing detergent on the brush (I like Dawn), and follow the steps below.

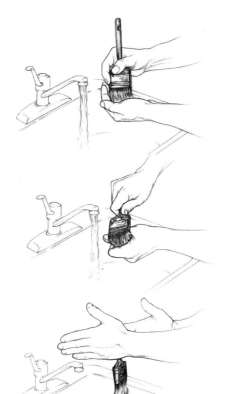

Cup your hand, and lather up the bristles with water. Swirl the bristles around vigorously.

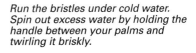

Rinse out the soap under warm water. Bend the bristles back to force out the finish at the base of the brush, near the ferrule. Repeat this until the bristles no longer feel slimy.

Run the bristles under cold water. Spin out excess water by holding the handle between your palms and twirling it briskly.

Straighten the bristles with a brush comb.

Wrap the bristles with paper (do not use newsprint; the ink will stain the bristles), and fold as shown. Lay the brush on a flat surface to dry. Don't store brushes in solvent for extended periods (more than four hours). The bristles will soften and lose resiliency.

Hardened finish
Soak the brush for four to six hours in a NMP (N-Methyl-Pyrrolidine) stripper such as Citristrip. Clean it as described above. Then, using a stiff wire brush, scrub the base of the bristles near the ferrule to remove the softened finish.

much as 100% of their weight in water, causing the brush to become soft and floppy in water-based finishes.

Modern manufacturing can now duplicate the natural flags of bristle. These flags are made by wire wheels that create a microscopic score along the entire length of the filament so that the tip will continue to split as it wears. Synthetic filaments also are less expensive and much easier to clean because they don't have microscopic pores of natural hair that trap finish.

Brush and bristle variations

Brushes for painting and varnishing are available in flat trim, rectangular chisel, oval chisel and touch-up. Flat-trim brushes (see the drawing on p. 70) are used for exterior painting; the blunt edge works the paint into the pores and crevices of the wood. I use these brushes for applying paste wood filler. The chisel edge on rectangular and oval brushes is used where precise control is needed, such as on moldings and edges. An oval profile has more bristles so it carries more finish (see the photo on p. 70). This is desirable for oil varnishes because the finish should flow on to minimize bubbles.

Touch-up brushes are assembled so that the tip ends in a round, fine point. These are the best brushes for detail work and painting fine grain lines in restoration work.

Buying a brush

Staining and general painting don't demand a great brush. But for applying finishes like varnish, which must be flowed on smoothly, a poorly made brush just can't do a good job. Be prepared to spend around $25 to $35 for a 1½-in. to 2½-in. China bristle brush of good quality.

When shopping for a brush, unwrap it. The bristles should feel soft at the tips and have spring in the overall length of the bundle. Examine the tips to make sure they have natural flags. Then pinch the whole thickness of bristles a little below the ferrule to see whether the fullness is the result of a lot of bristles (good) or a large divider (bad). Finally, fan back the brush with your hand. If the bristles come loose, don't buy the brush. And the color? The color of the bristles has no effect on performance.

Same width, more bristles. The oval chisel-edge brush (right) will hold more finish than the flat chisel-edge brush. This allows varnish to be applied in long, smooth strokes.

Synthetic bristles rival nature. Chinex bristle brushes (front) look, feel and work like the natural hog-bristle brush.

A HAND-RUBBED OIL FINISH

by Tom Wisshack

Achieving an open-pored look is as simple as eliminating all the intermediary sanding and jumping straight from plane or scraper to 600- or 1,000-grit sandpaper. More open-pored woods, such as the wenge in this tabletop, lend themselves better to this treatment than do cherry or maple.

Thomas Sheraton, the 18th-century English furniture designer, recommended making a paste of linseed oil and ground brick dust and rubbing it into mahogany with a piece of cork. The result, enhanced by innumerable polishings with beeswax over the years, is the beautiful patina we see on many treasured antiques.

Oil finishes still have much to offer today's craftsman. An oil finish will accentuate the grain, color and figure of the wood rather than obscure it, as many coats of a surface finish (such as varnish, shellac or lacquer) are prone to do. Additionally, an oil finish will never chip, peel, develop fisheye or orange peel. And dust contamination is not an issue with oil finishes, making them a good choice for the craftsman without a separate finishing space. If dust lands before the piece is dry, simply wiping it down with a soft, clean cloth takes care of the problem. Finally, and perhaps most importantly, because an oil finish penetrates and bonds with the wood, rather than forming a film atop the wood, renewing the finish is as simple as rubbing in some fresh oil.

As simple and beautiful as oil finishes are, however, it would be a mistake to view oil finishing as a quick, easy solution or a cover-up for bad workmanship. On the contrary, there is quite a lot of work involved in preparing a surface for an oil finish, and an oil finish will magnify any imperfections in the wood. Also, an oil finish is only moderately resistant to water and alcohol, so it may not be the best choice for a dining room or kitchen table, but for a piece of furniture subject to less spillage and daily wear, it may be ideal. For many craftsmen, the beautiful, rich patina that an oil finish develops over time far outweighs the care needed to maintain it. In this article, I'll discuss preparing for and finishing new furniture as well as rejuvenating previously oil-finished pieces.

Surface preparation

Someone once said that you could put used motor oil on a perfectly prepared wood surface and it would look good. As shocking as that may sound, the statement points out a fundamental truth: An oil finish is only as good as the surface to which it's applied. You may be able to get by with a less than per-fectly prepared wood surface if you plan to varnish or lacquer because these finishes form a relatively thick coating. But with an oil finish, any flaws in the unfinished surface will only become more evident when oiled, so you need to take extra care preparing the surface.

Some craftsmen prefer a handplaned or scraped surface to one that has been sanded a great deal. A surface finished by a cutting tool rather than sandpaper possesses a different tactile quality and will respond quite well to an oil finish. Most of us, however, find it necessary to sand at least a bit; how

fine a grit you stop at is largely a matter of personal taste. A surface that has been sanded to 1,000-grit will respond as well to an oil finish as one that has been handplaned only, but the characters of their surfaces will differ.

After planing or scraping to remove any mill marks or other imperfections from the wood's surface, you should raise the grain with a sponge or rag soaked in hot water. This will make any unseen flaws in the surface evident, so you can scrape or sand them out. It will also make your project easier to repair if it comes into contact with water after it's finished.

Homemade linseed-oil mixture rubs in best

Although there are a host of commercially available premixed oil finishes, I prefer to make my own. Call it part nostalgia, but it's the best oil finish I've used. I use this finish only on new furniture. If you're asked to restore an antique, you should seek the advice and expertise of a conservator before proceeding. Although eminently repairable, an oil finish is not removable save by sanding to bare wood.

I mix three parts boiled linseed oil (it must be boiled) to one part turpentine or high-quality mineral spirits and add a few drops of japan drier (generally available through commercial paint supply stores)—about two percent by volume. For the first coat, I warm the mixture in a

Burlap, rottenstone and the author's homemade linseed-oil mixture combine for a finish that's second to none. Although the paste formed by the rottenstone and oil mixture looks as though it would darken the wood, as long as there are no cracks, the paste will all come off.

I usually begin sanding with 220-grit wet/dry sandpaper on an orbital sander or hand-held sanding block. I follow up with 320, 400- and 600-grit paper, always sanding in long, straight strokes with the grain. A pine block faced with sheet cork (available from art-supply stores) will keep you from creating valleys as you would if you held the sandpaper in your hand; this is more important with the coarser grits because of their greater cutting effect. By the time you finish with the 400-grit, you'll start to see the wood grain and color come into focus. With the 600-grit, you're actually burnishing the surface. You may wish to use intermediate grits, or follow the 600-grit

with finer automotive sandpapers, but I find the above routine generally sufficient.

After attending to all flat surfaces, I take a piece of worn 600-grit paper and gently round any sharp edges and corners. This will prevent finishing rags from catching and will also give the piece of furniture a slightly used or worn look. If you wish to retain a more open-pored look, or would like hand-planing marks to be evident in the finished piece, skip straight from plane to 600- or 1,000-grit paper to polish the surface quite beautifully without filling all the pores (see the photo on p. 74).

It's important either to vacuum or to clean the surface thoroughly with com-

double boiler or electric glue pot, being extremely careful to avoid spilling any. I work a liberal amount onto one surface at a time using a natural bristle brush. Then I let the oil sit and soak into the wood for about 30 minutes. Next, I sprinkle the wood surface with a small amount of rottenstone and rub with burlap until a paste develops. I continue rubbing into the wood's surface for several minutes (see the photos on the facing page and above). Then I wipe all traces of oil and rottenstone off of the piece, using clean, dry rags. Remember that rags saturated with linseed oil are extremely flammable: submerge them in water immediately after use, or spread them flat outdoors to dry, and then be sure to put them in a closed garbage can outdoors at the end of the day.

I try to let the first coat dry in a well-ventilated, relatively warm area for about two weeks. If any oil beads appear on the surface during this time (they'll usually show up in the first couple of days), I wipe them off with a clean piece of terry-cloth towel. I apply the second coat more sparingly with a soft cotton cloth. After letting the oil soak in for about 15 minutes, I wipe off any oil remaining. I wipe until the rags come off the surface clean and dry and then give all surfaces a brisk rub. Two weeks later, I apply the third coat in the same fashion. If I'm going to apply a fourth or fifth coat, I'll wait another couple of weeks.

The drying time of this finish will vary tremendously depending on atmospheric conditions. The longer you can wait the better. It's possible to add more japan drier to the mixture to ensure drying, but the actual curing of an oil finish takes months and cannot be hastened chemically. Applying too many coats of oil in a short amount of time results in a greasy, slightly transparent tone. It's best to wait until the finish has begun to cure and form the beginnings of a patina before passing the piece on to a customer or gallery.

The hue of sun-bleached walnut suited the author, but the tabletop needed work. Using mineral oil and rottenstone, he rubbed out numerous minor scratches and scuffs, as well as gave the surface a new shimmer, without changing the color of the wood.

pressed air after each successive grade of sandpaper to avoid scratching the surface with particles left over from the previous, coarser grit. I also check the surface with a strong light between each sanding and again when I think I'm done. This will often reveal minor flaws I might otherwise have missed. The wood's surface, ready for oil, should have a sheen and be glass-smooth even before any finish is applied.

I like to let a piece of furniture sit for several weeks after preparing its surface and before I apply any oil. This time allows the surface to oxidize somewhat, giving it a head start on the rich color it will acquire with age. Cherry, for example, will look rather greasy and anemic and may have an unpleasant orangey tone if finished with oil right away. By letting the wood mature prior to finishing—even for just a couple of weeks—a richer tone results and the patina will build up more quickly. Not all woods respond to this waiting period, and not all craftsmen can afford to wait or are willing to do so. For me, the results are well worth it, and because I normally have several projects going at once, time isn't a problem.

Repairs and rejuvenation

An oil finish needs to be maintained. I'll refurbish one of my own pieces every couple of years, or sooner if it's damaged. To rejuvenate a surface that is intact (no scratches, water marks or abrasions), I simply rub my homemade oil finish into the surface for a couple of minutes and then remove all traces of oil with a dry rag. Finally, I rub the surface with another dry, clean rag until the surface has a satiny sheen.

If the surface is scratched or otherwise blemished, it's usually possible to remove the blemish by rubbing it out with a pad of 0000 steel wool soaked in the oil finish. Sprinkling a little rottenstone (a gray, abrasive powder much finer than pumice) onto the wood surface while rubbing will restore its original sheen. If you're removing a blemish from one area, in order to keep the same color and sheen over the whole piece, it's important that you not forget to rub the whole piece out. With each rubdown, the wood gets more beautiful and begins to form a patina. A table I made about ten years ago has had its top rubbed down about six times and is quite striking in appearance.

If a blemish doesn't respond to rubbing out with the steel wool, you may need to use wet/dry sandpaper with the oil solution. Although it depends on how deep the scratch is, as a rule, I don't use anything coarser than 320-grit for repairs. I use a sanding block (to prevent my fingers from digging into the wood) and follow the grain of the wood. Once I've removed the blemish, I work my way through the various grades of sandpaper until I have a perfect surface again, and I finish up with 0000 steel wool and rottenstone. I'm very careful not to sand too deeply because this would expose the underlying (nonoxidized) wood color, necessitating a much more extensive repair. Using the finest grade of sandpaper you can get by with will generally keep you out of trouble.

If you need to repair a piece of furniture but don't want to darken it, rub the piece down with mineral oil instead of a finishing oil. I have a walnut writing table that the sun had started to fade. I liked its color and wanted to retain it, but the tabletop needed some attention. Using the mineral oil just as I've used the homemade finish on other pieces (with a pad of fine steel wool and some rottenstone), I was able to repair the table without changing its color.

Choosing and applying oil

As I've tried to stress already, the kind of oil you use isn't nearly as important as the preparation prior to the actual finishing. I generally use a homemade oil finish (see the sidebar on pp. 76-77), but there are also a host of commercially available oil finishes. Danish oil finishes are among the most popular because they're simple to apply and the results are predictably successful.

Second in popularity to Danish oil finishes are tung oil finishes. The working properties of these finishes are similar to the Danish oil finishes, although tung oil gener-

ally cures faster and offers a bit more protection than most of the Danish oil products. (Keep in mind, however, that there is tremendous variability in formulation, drying time and working properties from one manufacturer to another. I've used tung oil finishes that have gone on like Mazola and stayed that way and others that started to tack up almost immediately upon application.) I find tung oil finishes too shiny, and in some cases, streaky for my tastes, especially with more than two coats, but a final rubdown with fine steel wool will generally both even out the finish and tone the gloss down to a satiny sheen.

My application procedure is similar for Danish oil and tung oil finishes. I brush on a first coat—liberally—and allow it to soak into the wood—about 10-15 minutes for Danish oil finishes but only 2-3 minutes for the tung oil finishes. Then I wipe up all oil remaining on the surface with a clean rag. I let this first coat dry for a few days (for either finish), and then I apply subsequent coats with a rag, wiping in a circular motion. Again, I eliminate all traces of oil remaining on the surface, using a clean, dry rag. Although there's no definite rule on how many coats you should apply, I usually give my pieces three to five coats. It's important to wait as long as possible between coats to avoid the greasy, hurried look that is characteristic of so many oil finishes.

Something to keep in mind, particularly with the more heavy-bodied oil finishes such as the tung oil finishes (although it's true to some degree with all oil finishes), is that the more coats you apply the more you lose the open-pored look. To retain this look on some of my contemporary pieces, I've applied only one coat of oil, and then followed that up a couple of weeks later with a coat of quality paste wax.

In instances where I want to finish a piece with oil, but a greater level of protection is required, I use Formby's Low Gloss Tung Oil Finish. The combination of tung oil and alkyd resins provides considerably more protection than most oil finishes, and the Formby's finish dries quickly and reliably.

TWO-DAY LUSTROUS OIL FINISH

by Sven Hanson

Building a lustrous oil finish in just two days is easy with the proper techniques. Preparation (sanding) and a smooth stroke, applying the right amount of finish, is the key.

Wood finishing is a repository for so much voodoo lore that some procedures should list chicken blood as an ingredient. But it can be done easily. I'll give you the basics and a plan for applying four successive coats of oil in 48 hours, using a clean rag to wipe clean oil onto a clean surface with just a bit of fine sanding between coats. The fourth coat comes out so smooth (see the photo below) that abrasion, wax or oil is necessary only to fine-tune the level of gloss desired. Really!

At first I made the natural mistake of assuming that if a surface looked smooth or felt smooth, it was ready for finishing. But when a board passes beneath the planer blade, the blade's rotation causes variations in cutting angle and height. Unless you smooth that cornrowed surface, it will reappear in your finish.

So you sand. But if you start with too rough a grit and don't get those sanding scratches out, you'll see swirls, especially if you apply an oil stain. Although they are hard to see, these scratches are visible in the unfinished or unstained piece.

Reducing the need to sand

Thorough sanding devours the hours, so I try to reduce the need by keeping sharp blades on my cutting tools and buying smooth, flat lumber. But understanding some sanding basics can really speed up things. Start with the finest grit that will do the job because a fresh piece of 100-grit, for example, cuts deeper than 150-grit. So I skip the 100-grit, except on bad tearout, and begin with fresh 150-grit paper, which also helps with reducing swirls.

My first sanding typically begins with a 150-grit belt on my belt sander. I like the belt sander because, in the hands of the skilled, you can create and maintain flat surfaces, and even a gorilla can't get swirls. Before I move on to 220-grit and finer orbital sanding, I thoroughly blow off the work surface and bench to remove the accumulated 150-grit. Then I sand with moderate pressure for the majority of the time and finish off every area at one-half pressure. This lets each grit of sandpaper remove some of its own scratches. To be sure that I sand evenly, I make a series of parallel pencil lines across the surface to be sanded. As the lines disappear, I can tell exactly where I have sanded, as shown in the photo on p. 83.

All woods improve in finishability with a light water rubdown before a final gentle sanding (see the photos on p. 82). This, by the way, is the first step in a blotch-free stain job. Serious smoothing calls for repeating this step until water no longer raises the grain. With some woods, the grain will continue to raise until hit with a first coat of

Water rubdown makes for smoother finish— The author raises the grain with water and then sands until the grain no longer raises. This procedure also helps reduce stain blotching.

A hair dryer speeds up the grain-raising process. After wetting the wood enough to darken the surface, a hair dryer quickly gets the surface sufficiently dry to sand.

finish, which raises the grain and locks it in place to be sheared off in the next sanding.

Oiling the wood

Protected by cheap vinyl lab gloves and working in a well-ventilated space, I begin oiling the wood. Many "oil" finishes are actually rubbing varnish. They're alkyd based, reduced with paint thinner, with lots of hardeners added. These "oils" offer the ease of application of oil combined with high solids for fast build and a hard drying finish that can be built up to a bright, protective surface. My favorite finish is Waterlox because it embodies all these features, is easy to use and usually is available at hardware stores.

I flow on a good wet coat with any absorptive rag. But don't use steel wool because it breaks down and darkens the pores of the wood. It also leaves behind steel fibers beneath the finish that can react with water and acids in wood, causing black splotches.

As the oil first goes on, I always spot a few flecks or streaks of glue. I immediately

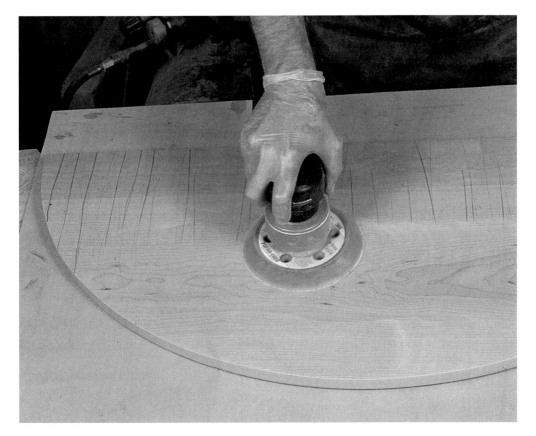

scrape them off with the back of a freshly sharpened chisel dragged across the surface like a scraper. (Hey! This isn't the top of a Steinway.) It usually blends right in, but when it doesn't, I sand the still-wet repair with a scrap of my usual 220-grit sandpaper.

Cleanliness doesn't matter for the first coat. The pores are full of dust, and some finishers actually sand the oily surface with wet-or-dry paper to make dust to fill the grain. Vacuuming just wastes time. Because of dust, fibers and the breaking of the finish film over the wood's pores, you can't create a sealed finish in one coat.

I lean or hang up the oil-covered workpiece, and when the first coat loses its gloss, I return to add more finish, usually in less than an hour. This wet-over-wet second coat needs only half the amount of finish as last time.

As with the next coat or two, I try to apply just enough finish to leave a temporary gloss without causing runs or drools. After 30 minutes, which can be shortened

by warm breezes and sunlight, I wipe the oil down, not off. By using a rag that contains heavy traces of oil in it, I avoid scouring the finish out of every pore. The rag leaves almost as much oil as it picks up. Think of it as "feathering off," like leveling the surface with a fine china-bristle brush. This is the essence of my fast-build system.

Hot air beats down beads

Oil stains and finishes have one nasty habit: beading up. You can apply finish and wipe it down to perfection, but when you return to see it in the morning, a constellation of tiny beads of soft finish has formed above the surface.

I beat the beads back by blasting them with hot air from my old hair dryer. The warm air lowers the viscosity of the oil, so it can penetrate better. The heat also speeds up the cross-linking process, so the oil cures faster. I hang up the work and check on it an hour later to wipe down any beads that might have formed in spite of my best

Building an oil finish—After the initial coat, the author lightly feathers on three succeeding coats in just two days to build a heavy, durable finish. Buffing with wax will help protect the finish after the oil has hardened.

efforts. It's then left to dry overnight. If you try to work it any sooner, you'll just soften the uncured oil and remove as much as you apply.

Day two, final coats

Next morning, I lightly sand with 220-grit or 320-grit sandpaper. This is the most important, and often the only, sanding required. Just a light pass will shear off the wild hairs, cut open the bubbles and knock down the few beads that popped up. Don't fear this step. If you give every square inch a light swipe or two with a folded piece of used paper, easy on the corners, then it's done.

Now you cleanliness freaks can sweep the shop, blow the dust off the walls and vacuum the work and the work surface. Don't worry about the dust in the air. The dust that will ruin your finish will come from the work surface, polluted finish, a dusty brush or rag, or fall out of your hair or sweatshirt.

After dusting myself off, I change into a fresh shirt and apron and put on a clean

dust mask. I blow off and then wipe the surface with a tack rag, making sure to clean out mortises, rabbets and around-the-corner areas where your once-clean rag can find fresh dust.

I filter the finish through a painter's filter or a clean muslin cloth. And I round up a well-worn but clean cotton sweat sock from my wife's sock drawer to use as an applicator.

I dampen the sock in clean finish and begin applying it to the difficult areas where two or more planes meet. I work from there to the outer, more visible, parts. Forget flooding and wiping off, and think of the process as brushing. I apply the amount of finish I want to remain on the work, spread it evenly, and then feather it all flat, as shown in the bottom photo. I stroke the molded edges and cross-grain parts first and finish wiping with the grain on the broad, flat parts. I try to complete the massage while the finish is still slightly liquid, so it can pull itself flat as it cures.

As I hang up the work, I'm often so amazed by the dazzling surface that I think three coats make a good enough finish. But no! Normally, after six hours of setting up, the finish has pulled tightly around microscopic swales and hillocks of wood pore and fiber. The flatness tells me I need another coat to fully protect the wood.

If you sand, use 320-grit paper. Barely caress every visible square inch of the work. Clean up and then apply the finish just like the last time. Take the afternoon off.

I promised you a lustrous finish in 48 hours, and you've got it. Four hours later, the piece will be ready to assemble or move with careful handling. I give it the fine-old-furniture feel by waiting a few days for the oil to fully cure. Then I apply a coat of paste wax with a superfine, non-abrasive pad and buff to a satin sheen.

Virtually every finish manufacturer specifies waiting before giving a final rub. A harder finish is less susceptible to damage and will rub out faster and more evenly. Wax topcoats require maintenance but provide extra protection from liquids and abrasion.

PADDING ON SHELLAC

by Jeff Jewitt

Padding shellac is a low-tech process that is perfectly suited to the professional and amateur finisher. The advantages of shellac are numerous. It is a nontoxic, Food and Drug Administration-approved natural resin. The carrier for shellac, ethanol, is relatively nontoxic (ethanol is the same kind of alcohol that's found in liquor), and the fumes are not unpleasant. Shellac dries quickly, so dust does not pose a great problem, and finishes can be done in two to three days.

Applying shellac by padding it on is an easy technique to master. I rub on a freshly

Padding on shellac doesn't require lots of fancy equipment. You can get a beautiful finish with a minimum of materials: shellac flakes, solvent, boiled linseed oil and wax. The author finished the tabletop in the background by padding on shellac.

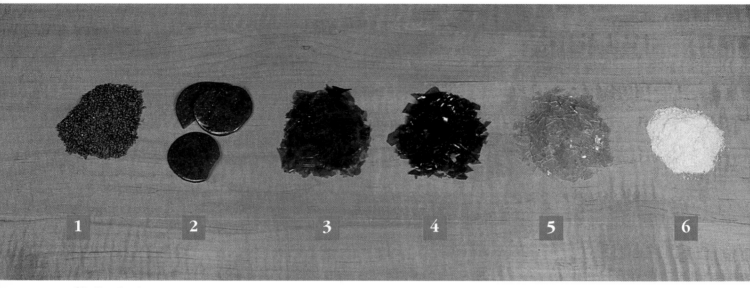

dissolved shellac solution over a sealer coat of oil, which increases the finish depth. I let each coat dry overnight and continue rubbing on shellac until I've achieved the desired depth and gloss I'm after. Shellac is a good-looking, durable finish that can easily be repaired if damaged. But because shellac can be dissolved by alcohol, this finish is not a good choice for a bar top.

The materials for padding shellac are inexpensive and easy to obtain through most finishing companies (see Sources of Supply on p. 89). They consist of shellac, denatured alcohol, padding cloth, a drying oil such as boiled linseed oil or tung oil, sandpaper and synthetic steel wool.

The materials

I prefer to make my own shellac solution of 2 lbs. of dry shellac flakes dissolved in a gallon of alcohol (a 2-lb. cut; for more on this, see the story at right). Using fresh shellac will help you avoid one of the classic complaints against shellac as a finish—it won't dry. Shellac is made up of organic acids that react with alcohol in a process called esterification. This gradual reaction produces esters, gummy substances that inhibit drying in old shellac.

Although it's possible to use premixed shellac, any liquid shellac older than six months should be tested for drying prob-

lems (Wm. Zinsser Co. makes shellac with a longer shelf life). To test shellac, place a drop or two on a piece of glass. If it's not dry to the touch in five minutes, don't use it. Premixed shellac is available only in orange or white (chemically bleached) varieties; there are more choices if you buy it in dry form (see the photo above). And if you mix your own shellac, you are guaranteed a fresh solution.

There are four alcohol solvents for shellac—methanol, ethanol, butanol and propanol. Methanol is an excellent solvent, but it's extremely poisonous. The fumes will pass through organic vapor respirators, so I avoid using methanol in my shop. Ethanol is far better because of its low toxicity. Butanol has an odor I find disagreeable, so I don't use it as the main solvent. I do add it occasionally to ethanol-reduced shellac as a retarder because butanol's higher molecular weight makes it evaporate slightly more slowly than ethanol. Propanol, the alcohol in rubbing alcohol, can be hard to get in chemically pure form. Don't use rubbing alcohol to dissolve shellac; it is 30% water and will cause problems in the shellac film.

An excellent product made specifically for reducing shellac is a Behlen product called Behkol (see Sources of Supply on p. 89), which is 95% anhydrous ethanol and 5% isobutanol. The isobutanol slows down the drying time slightly.

The best cloth for applying shellac is manufactured from bleached, 100% cotton and is sold as padding, trace or French polishing cloth. Whatever cloth you use, it should be clean, not dyed, lint-free and absorbent. Avoid old T-shirts or cheesecloth because of the lint. My favorite cloth comes in 12-in. squares and has a rumpled texture similar to surgical gauze, as shown in the photo on the facing page.

Use either boiled linseed oil or tung oil to seal the wood and to give greater depth to the finish (only a small amount is needed). I have not been able to discern a difference between the two under the shellac finish. Make sure the linseed oil is boiled, though, because raw linseed oil contains no driers and never really hardens.

Preparation

No finish can hide sloppy surface preparation. On new wood, I plane, scrape and sand to 220-grit on highly visible surfaces. I also do as much surface preparation as I can on the project before it's glued up. For new work, I'll even apply the oil and the first coat of shellac before assembling a project. Applying at least the first coat of shellac before the piece has been glued up makes it much easier to get an even finish, even in hard to reach places.

I generally tape off tenons and other joints so that oiling doesn't contaminate the wood. If the wood is to be colored, I use water-soluble dyes before the oil sealer coat. These dyes raise the grain, so I knock down the raised fibers with maroon synthetic steel

What's shellac and how is it used?

Shellac is derived from a natural resin secreted by a tiny insect called Laccifer lacca. This insect alights on certain trees indigenous to India and Thailand and feeds off sap in the twigs. The insects secrete a cocoon-type shell, which is harvested by workers shaking the tree branches. In this form, the resin is called sticklac and contains bits of twig, insect and other contaminants. The sticklac is then washed to remove impurities. At this point, it may be refined either by hand or machine. The next step up is buttonlac, which is processed in India. It is reddish-brown and is sold in 1-in.- to 2-in.-wide buttons.

Seedlac is another impure form of shellac and is processed further in India for better-quality lacs or exported to other countries for further refining. White shellac is made in the United States by Wm. Zinsser Co. from imported seedlac that's dewaxed and bleached by bubbling chlorine gas through it.

Shellac grading is complex because it is a product with wide commercial applications. But the most important characteristics for woodworkers are those based on color and wax content. The best grades of shellac for finishing have less than 1% wax and are light-amber in color. Wax in shellac decreases its moisture resistance and makes it less transparent.

The most common shellac is industry-graded as #1 orange, which usually is 4% wax and is a brownish-orange color. Dewaxed shellacs can range in color from a dark-golden brown

to a pale amber, as shown in the photo above. Fresh shellac is always better, so I mix my own, making just enough for the job at hand. For padding, I prefer a 2-lb. cut, which means 2 lbs. of shellac flakes dissolved in a gallon of alcohol. For most projects, a pint (¼ lb. of flakes in 1 pint of alcohol) is sufficient.

I mix shellac in a clean glass jar. Avoid metal cans because they will discolor the solution. Periodically shaking the jar prevents a jelly-like mass from forming at the bottom. Most shellacs take about a day to dissolve, so plan ahead. If it takes longer, the shellac may be bad. After dissolving in alcohol, lower-grade shellacs like buttonlac and seedlac always should be strained through a medium-mesh or fine-mesh filter to remove impurities.

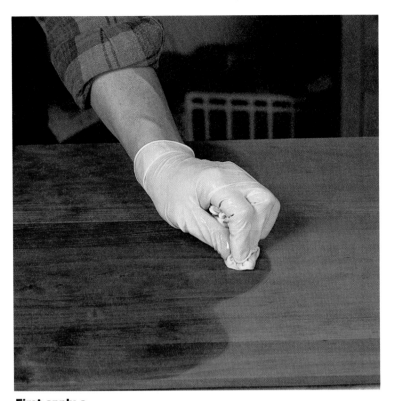

First apply a primer coat of oil for a deep finish. The author rubs in a light coat of oil, either boiled linseed or tung oil, to seal the wood. Shellac can be padded on after the oil has dried for several minutes.

wool (equivalent to 00 steel wool or 320-grit sandpaper) after the dye dries. I prefer synthetic steel wool because it's not as likely to cut through the dye on the edges. After the wood is smoothed down, you're ready for the first finishing step.

Oiling

Oil seals the wood and gives it greater depth. On refinished pieces, you can omit this step. Oils will accentuate the figure and deepen the color of wood, particularly curly maple and cherry. I have used a variety of oils, but I like linseed and tung oil the best. Apply just enough oil to make the surface of the wood look wet (about a thimbleful per square foot), as shown in the photo above. Do not flood the surface with oil. Apply the oil with a clean, soft cloth, and rub the surface briskly. It will penetrate quickly. After several minutes, begin applying the shellac.

Padding shellac

Fold the padding cloth into a rolled ball, as shown in the top photo. There should be no creases or seams on the pad bottom. Pour

about 1 oz. of alcohol into the pad and work it in. Then pour about $1/4$ oz. to $1/2$ oz. of a 2-lb. cut shellac into the bottom of the pad. I keep my shellac in round squeeze bottles to simplify dispensing into the pad. Use just a little; you shouldn't be able to squeeze shellac from the pad.

To apply the shellac, start at the top, right-hand edge of the board, and work across the board with the grain. Bring the pad down lightly, drag it across the board and right off the opposite edge, as shown in the drawing. Reverse directions, working back from left to right. Continue down the board, applying the shellac in alternating stripes. When you've reached the bottom, start again at the top; the board will be dry enough to repeat the process.

When the pad dries out, recharge it with more shellac. The amount of shellac you'll use depends on the size of the piece. A 24-sq.-in. piece should take about 10 or 15 minutes and will use three or four charges of shellac. On tops, do the edges first, and then continue the same sequence as above. If there is a complex molded edge, make the pad conform to the shape of the molding. The other parts of the piece (aprons, legs and sides) get the same padding coat of shellac. When the board is tacky and the pad starts to stick, stop. Store the pad in a jar with a screw-type lid.

The first application of shellac should be dry enough to scuff-sand in approximately 1 hour. Using 320-grit, stearated sandpaper (aluminum oxide mixed with zinc stearate as a lubricant), lightly scuff-sand the surface. Scuff-sanding is applying just enough pressure to barely scratch the surface. After this, smooth out the surface with maroon synthetic steel wool. Then apply shellac to the other sides of all surfaces, such as the undersides of tops and the insides of carcases in the same way you did on the top.

When this coat of shellac is dry, after about an hour, scuff-sand and rub these surfaces with synthetic steel wool. After the first coat of finish has been applied, it's time to glue the project together. Be careful to avoid excess glue, and make sure that clamps are properly padded. If any glue squeezes

After the oil dries for a few minutes, charge the pad with a squeeze bottle to get just the right amount of shellac. The pad should be a lint-free cloth folded so that there are no wrinkles or seams on the bottom of the pad.

out, you can pull it off like scotch tape after 30 minutes to an hour. Don't let the glue dry completely, it may pull off the finish when you try to remove it.

The next day, once the piece is glued up, the finishing sequence is repeated. The pad should glide easily over the surface, and you should have an even coat of shellac on the surface. As the pad starts to dry out, you can switch from polishing in a stripe pattern to a circular pattern or a series of figure eights to get even coverage on the board. Stop when the finish is tacky and the pad sticks. At this point, the surface should have an even shine, indicating a surface build of shellac. Put the pad back in the jar, and let the finish dry overnight.

The next day, examine the finish. You should have an even coating of finish on the surface. If you are working with open-pored woods like walnut or mahogany, you'll see crisp outlines to the open pores. This level of finish is appealing to some. If so, you can stop applying shellac; simply go on to the rubbing-out stage, which I'll explain in a minute, and you're done.

For surfaces that will receive a lot of wear and tear, you may want to apply several

SOURCES OF SUPPLY

The following companies manufacture or supply dry shellac flakes in various grades, padding cloth, alcohol solvents, oil and other finishing products.

H. Behlen & Bros.
Route 30 N.
Amsterdam, NY 12010
(518) 843-1380

Garrett Wade Co., Inc.
161 Ave. of the Americas
New York, NY 10013
(800) 221-2942

Homestead Finishing Products
11929 Abbey Rd., Unit G
North Royalton, OH
    44133-2677
(216) 582-8929

Olde Mill Cabinet Shop
1660 Camp Betty
    Washington Rd.
York, PA 17402
(717) 755-8884

Padding shellac

Shellac is padded on with the grain from edge to edge in an alternating pattern until the piece has been covered. The stroke should start off the edge of the board, continue across the board and off the opposite side. Stop when the finish becomes tacky, and the pad begins to stick.

more coats for maximum protection. If so, repeat the procedure until you've built up the finish to the film thickness that you want, allowing each coat to dry overnight. You don't gain any added protection after four or five applications, but there is an aesthetic difference. After the final padding application, let the project dry for several days before rubbing it out.

Rubbing out

Rubbing out the shellac finish results in a smoother, better-looking surface. The beauty of the padding application is that there are no brush marks or other surface irregularities to level, so this step usually goes quickly. The first step is to level the surface of the finish with 400-grit, wet-or-dry silicon carbide finishing paper. Then switch to 0000 steel wool, squirting mineral spirits onto the pad and dipping it into a can of paste wax.

I prefer steel wool for rubbing out because it has a better bite and leaves a better-looking finish. My favorite wax is Behlen's Blue Label paste wax, available in brown for darker finishes and natural for lighter finishes. Working with the grain, I bear down fairly hard with the steel wool and rub the wax on the surface. I wait until it begins to haze, wipe off the excess and buff to a satiny sheen. If a higher gloss is desired, rub the surface with rottenstone mixed with mineral spirits before waxing.

Maintenance

If the piece is not subjected to a lot of wear and tear, a yearly re-waxing keeps it looking great. For tables, chairs and other high-wear items, you can rejuvenate the finish by removing the wax with mineral spirits and rubbing with maroon synthetic steel wool. Then apply a light coat of shellac, let dry and re-wax.

PADDING LACQUER

by Mario Rodriguez

For me, French polishing is the finish of choice for the very finest furniture. When done well, a French polish has a soft but brilliant glow that brings out all the depth and color of the wood without the heavy buildup generally associated with a high-gloss finish. No other finish even comes close.

I've taught French polishing for years, and for beginners, it can be a nerve-racking juggling act. The ingredients of a French polish—shellac, oil and pumice—must be applied at the right time and in the proper amounts. The addition of each can improve the finish dramatically—or destroy it. Padding lacquer is an amazing one-step mixture of dissolved shellac, lubricants and nitrocellulose resins. It produces a surface virtually identical to that of a traditional

French polish, without the risks. It still requires a lot of elbow grease, but because it's a premixed formula, you can concentrate on applying it and not worry about maintaining a delicate balance of ingredients. There are several brands of padding lacquers from which to choose (see Sources of Supply on p. 95). I haven't found significant differences among them.

In addition to being convenient and easy to apply, padding lacquer dries quickly, so you don't need a special finishing room. It can even be applied on-site, eliminating the need to bring a piece of furniture back to the shop for finish repairs. And because shellac is the primary ingredient in a padding lacquer, it can be applied over other finishes. Finally, padding lacquer has a

**1. Scrape the
surface** until it's
flat and even in
appearance.

**2. Sand with the
grain** using 220-
and then 320-grit
sandpaper.

variable sheen. The more or less sanding you do will increase or decrease its gloss.

Surface preparation

For more formal furniture pieces, which generally look best with a high-gloss finish like a French polish, I scrape the wood until I have a fairly flat, uniform surface (see the photo at left above). Then I sand with 220-grit and 320-grit sandpaper (see the photo at right above).

After wiping the surface with a dry rag, I wash it down with denatured alcohol. This raises the grain slightly and allows me to see sanding scratches and any other flaws (see the top photo on the facing page). If I want to fill the pores slightly for a smoother finish, I wet-sand with worn 320-grit wet-or-dry sandpaper and denatured alcohol. If I want a glass-smooth, nonporous finish, I use a filler (for more on this, see the box on p. 94). For a moderately porous, more natural-looking finish, just dry-sand with 320- and 400-grit sandpapers once the denatured alcohol has dried.

Applying padding lacquer

When using padding lacquer, all you need is a 6-in. square of lint-free cotton. Old T-shirt scraps work great. Just make sure

that there aren't any creases or seams in the center of the pad because they can mar your finish.

I pour a small amount of padding lacquer into the center of my cloth and let it soak in a few seconds. Then with a small, circular motion, I begin to rub the polish vigorously into the surface (see the center left photo on the facing page). Initially, the surface will haze and the cloth will drag a little, but with firm, steady pressure, an attractive shine will quickly start to appear. As I move from one small area to another, I carefully overlap my applications for uniform coverage (see the center right photo on the facing page).

A second coat can be applied almost immediately. As you build up the polish, though, you should extend the time between coats for the best results. When I get to my fourth and fifth coats, I usually wait between 12 and 24 hours.

Feathering out the finish

Even with very careful application, some areas will have more of a sheen than others, and the overall surface may look splotchy. You'll want to go over duller areas and make the surface as uniform as possible.

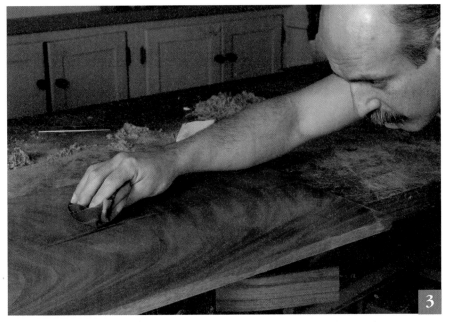

3. Check for sanding scratches and other flaws by flooding the surface with denatured alcohol. This also raises the grain slightly, so follow up by sanding with 320- and then 400-grit sandpaper.

4. Quick, circular motions bring up a shine. Move the pad in tight circles in a small area, applying a good deal of pressure. The surface will be hazy at first, but after just a minute or so, a shine will start to come up. Apply less pressure as the shine increases.

5. Work just a few square inches at a time, blending adjacent areas. Apply more pressure on unfinished areas.

6. Polish the whole surface lightly. Take a clean rag, apply just a little padding lacquer and rub very lightly. The rag should just skate across the surface. Do this until the whole surface has a uniform sheen.

Pore filler gives a glass-smooth surface

Pour it on, smear it around. You don't have to be fussy when applying wood filler—just fill all the pores. Move the rag around; then use a scraper.

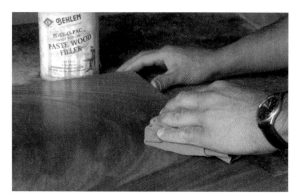

Filled pores, satin sheen—Paste wood filler dries to a satin sheen even before padding lacquer is applied. The filler dries rock-hard, so wipe the surface clean.

Like a mirror—With its pores filled, this crotch mahogany panel takes on a finish that's a dead-ringer for French polish—a awarm but brilliant sheen.

In traditional French polishing, pumice helps fill the pores in the surface.

Padding lacquer has no pumice, so the pores don't get filled appreciably, except by the padding lacquer itself. The result, depending on how much sanding you've done, is a relatively open-pored surface.

To get a glassy-looking, non-porous surface with padding lacquer, I use Behlen's pore-filling compound called Pore-O-Pac paste wood filler (see Sources of Supply on the facing page). Pore-O-Pac is available in six shades.

Applying the filler couldn't be easier. I pour some on the surface I'm going to polish and wipe it all around with a rag (see the top photo at left).

Then I use a scraper like a squeegee, moving the filler across the wood in all directions. This works the filler into the pores.

I let the filler remain on the surface between 30 minutes and one hour before wiping it off. This filler dries rock-hard, so it's important to clean the scraper and the surface you're filling. Otherwise, it will take a belt sander to remove it. I use a clean rag and keep wiping until the rag comes off the surface without any residue.

I wait 24 hours for the surface to dry, and then I fine-sand with 320- and 400-grit sandpaper. After sanding, I wipe down the surface with a rag soaked in denatured alcohol.

I let the surface dry and start applying the padding lacquer. A brilliant gloss will start to come up almost immediately (see the bottom photo at left).

Then put a small amount of padding lacquer on a clean rag, and apply it over the entire surface, using a broad, circular motion. Bring the cloth just barely into contact with the work surface—almost glancing over it. This will eliminate any small streaks or blotches and leave a consistently brilliant, thin film (see the bottom photo on p. 93).

Repairing mistakes

As easy as padding lacquer is to use, I do run into small problems from time to time. These problems usually appear as rough crater-like patches. If they're not too severe, I simply pad over them. The application of new material usually will soften the area and vigorous rubbing will level it out. If this doesn't do the trick, I'll let the panel dry overnight, scrape or sand the damaged area flush the next day and then repolish. After a coat or two, blemishes will disappear completely.

Finishing on the lathe

I often use padding lacquer on lathe-turned objects, including table pedestals, spindles, cabinet knobs and tool handles. Here the application is even easier. Sand to 320-grit with the object spinning on the lathe. Then raise the grain with alcohol, and sand again with 320- and then 400-grit paper. You can apply the padding lacquer a little more heavily on the lathe, but don't use so much that it's spraying off the workpiece. Use gentle pressure on the rotating workpiece, and watch an incredible gloss develop.

MAKING A CASE
FOR VARNISH

by Frank Pollaro

I'd just finished the most complex piece I had ever attempted, the reproduction of a desk by Emile-Jacques Ruhlmann, the greatest of the Art Deco furniture designers. The curvaceous desk, veneered in amboyna burl and shagreen, or sting-ray skin, had taken me more than 300 hours to complete. The original had been French polished, but I wanted to provide my reproduction with more protection than shellac affords while giving it the same clarity and brilliance.

I asked Frank Klausz, a friend and fellow woodworker, what he recommended, and he suggested that I use varnish. I experimented on scrap boards until I was satisfied with the results. And then I varnished the desk. It was the perfect finish with all the depth, clarity and brilliance I had hoped for.

Now varnish is the standard finish for all my fine work (see the photo at left). I've experimented with a number of varnishes and brushes and refined my technique. Now I can brush on a finish that looks as though it has been sprayed.

Understanding varnish

A properly applied varnish finish is glass smooth, hard and resistant to most household chemicals, foods and drinks. It also has a warm, amber glow. That makes it best suited for darker woods, unless you want to add warmth to a light wood, such as maple or ash. Regardless of the choice of wood, a well-polished varnish surface will turn heads.

Varnish must be rubbed out

About the only downside to using varnish is that you have to rub out and polish the finish if you want a blemish-free surface.

Always use a good brush. Look for a thick, firm brush with fine bristles, like this badger brush.

Brush on the varnish in long, smooth strokes.
On a surface with a single or a predominant grain direction (unlike this sunburst veneer pattern), start by applying the finish across the grain. The first coat of varnish should be a 50/50 solution of varnish and solvent.

Brush out the varnish at 90° to the direction you laid it on, usually with the grain. Use a light touch. Just skim across the surface without exerting any downward pressure.

Varnish is oil based, so it takes far longer to dry than lacquer or shellac. Lacquer thinner and denatured alcohol evaporate in minutes, leaving a hard, dry finish behind. Varnish can stay tacky for hours, vulnerable to anything in the air, whether that's dust or a wandering fly. So it's important to apply varnish in as clean an atmosphere as possible.

Depending on the style and function of the piece of furniture I'm finishing, as well as the client's tastes, I may polish it only to a satiny gloss, or I may take it all the way to a high gloss. Either way, though, it's not nearly as time-consuming as a lot of woodworkers think it is. Even a very large dining table won't take more than an afternoon to rub out and polish.

You must sand between coats
The other major difference between varnish and lacquer is that you cannot reactivate dried varnish with a fresh coat or with a solvent. With lacquer, every time you apply a new coat of lacquer, you effectively melt it into previous coats, creating what amounts to a single, thick coat. With varnish, you're building up a finish one layer at a time. Each new coat should bond mechanically to the one below it by gripping the scratches in the surface. For this reason, it's absolutely essential to sand between coats until there are no shiny, low spots.

One final detail about the varnish itself. Always use a high-quality product. It will brush on and flow out much better than cheaper stuff. I've settled on Behlen's Rockhard Tabletop varnish (distributed through Garrett Wade; 800-221-2942 and Woodworker's Supply; 800-645-9292). It's the best varnish I've found, and it dries the hardest, so it rubs out better than any other.

A good brush is the key
The single most important thing you can do to achieve a great varnish finish is to start with a good brush. They aren't cheap—expect to spend between $30 and $60 for a 3-in. brush. My first varnish brush was a badger brush from Behlen's (see the photo above), which I still use. It's a good value at $30 or so. But I discovered another brush last year that I like even better. It's made in Germany from the inner ear hair of oxen and is imported by Kremer Pigments (228 Elizabeth St., New York, N.Y. 10012; 212-219-2394). The brush, listed simply as the Pi72, costs nearly $60. But it has very fine bristles, which leave virtually no brush marks in the finish surface.

Whichever brush you decide to use should be thick, firm and made with fine, natural bristles. This will allow the brush to hold a good amount of varnish and distribute it evenly on the surface. A thin, skimpy brush won't hold enough varnish. A limp

Rub out nubs or bumps with 1,200-grit paper wrapped around a wooden block. Water, naphtha or mineral spirits may be used to lubricate the surface.

Use a rubber squeegee to clear slurry. The 1,200-grit paper works slowly, so keep rubbing and clearing the slurry until all the high spots are gone.

brush won't move the varnish around, and coarse bristles can leave marks in the finish. If you're going to use varnish, do yourself a favor and buy a good brush.

Brushing it on

The best place to varnish a piece of furniture is in a small, dust-free room with the windows closed. Few of us have that luxury, though. To reduce the number of little dust specks settling on the wet varnish, I often spray a mist of water in the air, on the ceiling and on the floor just before getting started. Try not to get any water on the piece you're about to finish. Don't get too worked up about dust, though, because any small bumps will be sanded off after each coat has dried.

I cut the first coat of varnish 50% with thinner and add a few drops of Behlen's Fish Eye Flo-Out. This is essentially just silicone, but it enhances the flow of the varnish, eliminates the likelihood of fisheyes and improves the scratch resistance and glossiness of the finish.

Brush technique is important with varnish. The object is to apply a thin, even coat. If you put on too much varnish, it will skin over and the varnish under the skin will never dry. If you use too little varnish, you'll have a hard time moving it around, and it will not flow out. With a little practice, though, the whole process will become second nature.

I find it helpful to let the brush soak in the varnish for a minute or two, so it can absorb some of the finish. Then I apply the first coat, brushing all the way across the table in long, smooth strokes (see the left photo on the facing page). After covering the table with varnish, I quickly brush over the varnish I've just applied, but at 90° to the original direction and with a much lighter touch (see the right photo on the facing page). Each coat is applied in the same way. On a piece of furniture with a predominant grain direction, I apply the varnish first across the grain and then brush it with the grain. You have to move quickly because even though the varnish will stay tacky for hours, it will start to set up after just a few minutes. You'll probably see brush marks, or striations, in the surface at first, but after 15 minutes or so, they'll level out.

I let this first coat dry for at least 24 hours and then sand it out with a random-orbit sander and a 220-grit disc. This gives the surface some tooth for the next coat to bind to. After sanding, I wipe down the surface with a tack cloth before applying the next coat.

I brush on the second coat, cut with 25% thinner and then wait another 24 hours for the coat to dry before sanding it. For a tabletop like this one, I'll apply four or five coats, allowing 24 hours between each coat and 72 hours after the last coat before starting to rub out the finish. The third and

Polish the finish with a power buffer and automotive glaze. Once you've sanded out all the nubs and bumps and gotten the surface flat, 10 minutes of power buffing will take the finish to a high-gloss shine.

subsequent coats are full-strength varnish. Four coats are usually enough, but I've applied as many as eight. If you want the surface to be completely smooth and nonporous, keep applying coats until there are no pores showing after you've sanded with the 220-grit paper. Then just one final coat should do it.

Rub out and polish the finish

When you're happy with the last coat and have given it at least 72 hours to dry (a week would be better), it's time to rub out the finish. For a satin finish, I just sand with 600-grit paper and polish with 0000 steel wool lubricated with Behlen's Wool-Lube. Then I rub down the surface with a clean cloth, and I'm done.

For a high-gloss finish, I used to wet-sand from 600-grit to 1,000-, 1,200- and, finally, 1,500-grit paper. Now I start and end my sanding with 1,200-grit paper (available at most auto-body supply shops). The advantage of working your way through the grits is that the rubbing out takes less time and the result is likely to be slightly flatter because you're starting with a more aggressive abrasive. The reason I stopped doing it is that I always found myself trying to eliminate a scratch or two from one of the coarser grits that only became apparent after I'd gotten to the 1,500-grit. I'd have to go through the whole routine again, losing any time I had saved.

To take down all the nubs or bumps in the surface of the finish caused by dust or

For porous woods, fill in the grain

On very open-grained woods, such as burls, I collect all of the sawdust from my final dry-sanding (220-grit) in a jar. I mix this sawdust with full-strength varnish (see the top two photos at right). I hone a square edge on a 2-in. putty knife and use it to apply this paste to the raw wood in place of the 50% dilution I normally use for the first coat.

I lay this paste down in one direction and spread it perpendicularly. I fill the voids, imperfections and pores (see the bottom photo at right), being careful not to scratch the surface. After 24 hours, I sand with 220-grit to reveal a glass-smooth surface. Two more full-strength coats of varnish and I'm ready to rub out and polish the finish.

Mix full-strength varnish and 220-grit sanding dust until it has the consistency of molasses.

Work mixture into the grain. Apply it in one direction, and then work it into pores crosswise. Try to create a smooth surface.

other debris, I wrap the sandpaper around a wooden block (see the left photo on p. 99). I've used naphtha, mineral spirits and water as wetting agents. For this table, I used water with a little Behlen's Wool-Lube in it to make things more slippery. A little rubber squeegee helps to clear away the slurry, so you can check to see if a bump is gone or if you have more sanding to do (see the right photo on p. 99). The auto-body supply dealer I do business with gives me these squeegees.

After I've sanded out all of the nubs and bumps, I swap the wooden block for a cork block and give the whole table an even sanding, trying to get it as flat as possible. It's important to take down any high spots after each coat. If you let these spots build up, you could sand through one coat into another. This shows up as a visible ring between the two coats, and the only way to fix it is to sand off the whole topcoat and apply it again.

Pay special attention to the edges, where the varnish can build up a little ridge. You

can judge how flat the finish is by looking at the reflection of a light on the table. If it looks like it's reflecting off the surface of a wind-swept pond, then you have some more sanding to do. If it's relatively undistorted, you're in good shape.

To complete the gloss finish, I apply Meguiar's Mirror Glaze #1 (an automotive rubbing compound), buff it out and wipe it off. (For the closest dealer, call Meguiar's at 800-854-8073.) It's important to get the surface completely clean because any residue from the #1 compound will scratch the surface when you go to the next finer compound. I follow the Meguiar's #1 with the #3 compound, using a different buffing wheel—again, so the residue from the coarser compound doesn't undo what I'm trying to accomplish (see the photo on the facing page). After buffing with the #3 compound, I wipe off the table with a clean rag. The surface will shine like a mirror.

RUBBING OUT
A FINISH

by Jeff Jewitt

A flawless finish— Hand-rubbing eliminates surface defects, which can mar even carefully applied film-forming finishes. The author's three-step approach includes sanding out surface imperfections, leveling the surface and then polishing to a uniform sheen.

A cured finish rarely looks or feels blemish-free, no matter how carefully you applied it. Bubbles, dust and debris can lodge in the finish as it dries and are especially noticeable on gloss finishes. Rubbing out a finish eliminates blemishes, so it should be the last step in finishing any piece of furniture. Surprisingly, few finishers I know do it. No doubt, some fear having to abrade a finish film that's only thousandths of an inch thick. Taken in steps, though, rubbing out a finish need not be a terrifying process.

Any film-forming finish will rub out: hard finishes, like nitrocellulose lacquer, flexible ones, like polyurethane, and even waterborne finishes, which can be challenging to polish (see the top photo on p. 104).

Rubbing out is a three-step process of removing imperfections, leveling the surface and polishing to a consistent sheen. I always use gloss finish because it can be buffed. Satin-formulated finishes contain silica flatteners that impede light reflection, so they can't be rubbed out to a gloss (for more on this, see the sidebar on p. 106).

Prepare the work surface and the finish

If you're finishing an open-grain wood like mahogany or oak, fill the pores first. If you don't, light-colored abrasives will lodge in the pores and will be visible. If your wood is textured (from a handplane, for example), sand or scrape the surface flat before you put on the finish. Otherwise, you risk rubbing through high spots and exposing the stain layer or bare wood. In situations where you can't flatten the surface (inlaid furniture and hand-tooled antiques, for example), you'll have to rub gently with steel wool. The wool acts like a cushion, so it's not as likely to shear off the high areas.

Rubbing out removes finish, so be sure to start with a thick coating. Solvent-release finishes, like shellac, lacquer and some waterbornes, fuse into a single film once they're applied. With these finishes, I generally apply six coats.

By contrast, coats of reactive finishes, like oil varnish and polyurethane, do not melt into one another. If you rub too much, you'll go through the top layer (see the bottom photo on p. 104). Most reactive finishes have a higher solids content, so I usually apply only three coats, and I make sure that the last coat is not thinned.

Fully cured finishes buff up better and faster than finishes that aren't. Shellacs, lacquers and two-part finishes should cure at least a week. Oil-based varnishes and polyurethanes should cure at least two weeks. If the finish is gummy and loads up your sandpaper, let it dry longer.

Keep in mind, too, that soft or flexible finishes do not rub out as easily or to as great a shine as hard, brittle ones. It's like the difference between polishing the sole of your shoe compared to polishing brass.

Sand imperfections

The first step of rubbing out is using abrasive papers to dry-sand or wet-sand defects from the cured finish. If the finish is in good shape, you can skip this step. Dry-sanding can be dusty and tedious, but at least you can tell what you're doing to the surface. Stearated aluminum-oxide paper works well for this, though it will clog fairly quickly on hard finishes like lacquer and shellac. Several new papers are available that have precise, uniform grit sizes. Although they are more costly, 3M's Microfinishing paper and Meguiar's Finesse paper (available at most auto-supply stores) are worth trying. They cut much more efficiently.

Wet-sanding is fast, but the slurry can give you a false sense of finish thickness. It's easy to sand through to the sealer or to the color coats. Wipe the surface, and check your progress regularly. Rub very lightly near the edges of flat horizontal surfaces where the finish is likely to be thinner.

I prefer wet-sanding with traditional wet-or-dry paper, and I use water with a dash of dishwashing soap as a lubricant. I usually sand the finish with 400-grit, but I'll go to 320-grit if there are big hunks of debris to remove. I wrap the paper around a cork block and sand enough to knock down the tops of dust pimples, so they're even with the rest of the finish. On curved surfaces, I use my hand as a backer. When the imperfections are gone, the surface is speckled with alternating dull and shiny spots.

How you apply it doesn't matter.
These three panels, two mahogany and one walnut, are buffed to a gloss. From left, the finishes are sprayed-on nitrocellulose lacquer, brushed-on rubbing varnish and waterborne acrylic lacquer, which also was brushed on.

Witness lines show rub-throughs—Separate coats of reactive finishes, like water-based polyurethane, remain distinct, so be careful not to rub into an under layer. Witness lines, which look like feathery rings on this piece of birch, are the result.

Level with finer abrasive papers
Leveling establishes a consistent scratch pattern on the finish. I level with 600-grit, but if the surface is rough with brush marks or orange peel, I start with 400- or 320-grit. Wrap a clean sheet of wet-or-dry paper around your block, and squirt some soapy water onto the surface (I use a plant mister). Sand all the edges first. Don't worry about the grain direction.

Next, work the center of the board in manageable sections using a crosshatch pattern (see the photo at left on the facing page). Rubbing from opposing 45° angles ensures complete leveling. Now rub with the grain. As you sand, keep exposing clean, fresh grit. Change to new paper often. Finishes can gum up and clog paper quickly.

Wipe away the slurry with a rubber squeegee, and look for shiny spots under backlighting. Squirt on more water, and rework areas that are still shiny, but don't overdo it. To avoid making hollows, feather each area into the rest of the surface. You can leave very small shiny areas because they won't be too visible once the whole surface is buffed. Rub shiny spots near the edges with dry steel wool until they're dull like the rest. When you're satisfied, switch to the next finer grit and repeat. Continue on to 600-grit.

Polish with steel wool or powdered abrasives
The last step is polishing, and you have a choice here: satin or gloss. When I want a satin finish, I rub out with 0000 steel wool or synthetic steel wool lubricated with soapy water and Behlen's Wool-Lube (or paste wax thinned with mineral spirits). When I want a gloss finish, I use traditional powdered abrasives—pumice (powdered volcanic glass) and rottenstone (powdered decomposed limestone) mixed with water or oil. Pumice is sold in grades from 1F (coarse) to 4F (fine). Rottenstone is finer than pumice and is sold in one grade.

For a satin sheen
Squirt some soapy water on the finish, and apply a generous dab of Wool-Lube to a wad of steel wool (unravel it, and fold it into quarters to make it last longer). You can also

Satin finish
After leveling, use steel wool. For the finish to have a satin sheen, the author uses 0000 steel wool, soapy water and either thinned paste wax or Behlen's Wool-Lube to polish the surface.

Level with abrasive papers. Wet-or-dry paper backed by a cork block makes a flat abrasive pad for leveling. Keep the surface wet while rubbing crisscross over the middle of the panel. Squeegee off the slurry to check your progress.

Gloss finish
For a gloss look, use pumice and rottenstone. Sprinkle on pumice and then wet and rub the slurry over the whole surface. Wipe this off, and follow with rottenstone.

use a gray Scotch-Brite nylon pad or equivalent grade of synthetic steel wool. Rubbing back and forth with the grain, make nine or 10 slightly overlapping passes. Use two hands for firm, steady pressure (see the photo on p. 102). Wipe away the slurry to make sure you're creating a uniform scratch pattern. You may have to let the slurry dry to see if you've got it right. If you want a silky feel to the surface, let the slurry dry on the surface, and then buff it off just like wax. When backlighted, a satin surface should look like brushed metal.

For a gloss sheen

Skip the above step, and continue wet-sanding up to at least 800-grit. I take it to 1,200-grit. Now sprinkle on some 4F pumice (see the bottom right photo above), and wet it with water or rubbing oil. Wad up a clean, dry cotton cloth and, working in whatever direction you want, polish every

square inch of the surface. Use lots of pressure, and replenish the pumice and water as the slurry dries. Let it haze over, and then wipe it off with a damp rag. Switch to rottenstone and do the same.

Turnings, carvings and moldings

To rub out intricate surfaces, like turned legs or carved aprons, polish with 0000 steel wool and thinned wax. The finish in these areas is just too thin to polish to a gloss. Don't rub too hard, or you'll cut through the finish on sharp details. To avoid a light-colored wax residue on dark finishes, use dark-colored wax. For moldings, wrap some 600-grit wet-or-dry paper around a sanding block that's shaped to match the convex or concave curve of the molding. Rub with steel wool and wax. When the wax is dry, it can be buffed so it approximates the sheen of the rest of the piece.

Sheen is a measurement of reflection

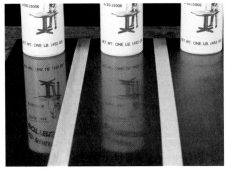

Finish manufacturers measure sheen using a gloss meter, a device that reads how much light is reflected off a surface. Tests for finishes containing flatteners measure light reflectance at 60°. When the angle of incidence (incoming light) equals the angle of reflection (outgoing light) and at least 80% of the light is reflected, the sheen is considered gloss (see the top drawing below). Semigloss finishes reflect between 70% and 80% of the light; satin finishes reflect 35% to 70%; flat, matte and eggshell finishes reflect 15% to 35%.

But you don't need a gloss meter. A simple visual test can be used instead (see the photo above). A finish that gives a clear reflection, with clean, distinct outlines is gloss. If the reflected image is readable, but fuzzy, the sheen is semigloss or satin. When little or no light is reflected or the reflection is no longer distinguishable, the sheen is eggshell or flat.

Scratches from polishing influence light the same way that flatteners added to the finish do. Both the size and the depth of the scratches affect the interference pattern, or scattering, of the light. As a general rule, the scratch pattern left from 400-grit abrasive paper produces a dull or flat luster. Finishes abraded to 1,000-grit appear satin. Those scratched with 1,200-grit and higher produce sheens ranging from semi-gloss to gloss.

Interestingly, when you lower your viewing angle on any surface, the sheen appears more glossy. That's because you're seeing less diffused light.

How sheen affects light

Sheen is a measurement of how much light reflects off a surface. Flatteners or scratches on a finish diffuse light, which is why satin and flat finishes reflect less light than gloss ones.

Smooth surface

When angle A = angle B and at least 80% of the reflected light reaches your eye, the finish surface is a gloss.

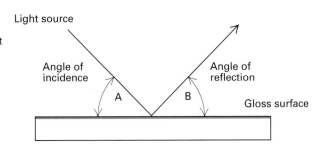

Rough surface

Scratches on a finish diffuse light. Finishes rubbed out to flat or satin diffuse more light than gloss finishes. The diffused light makes a reflected image less distinct because less light reaches your eye. Generally, the finer and more uniform the scratch pattern from polishing, the glossier the sheen.

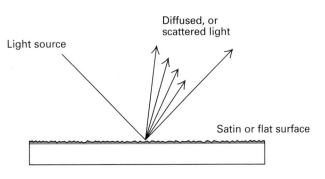

REJUVENATING WITH WAX

by Tom Wisshack

A coat of paste wax is probably the simplest and safest way of rejuvenating the surface on a piece of old furniture. In the 18th and 19th centuries, varnished, French-polished and oil-finished pieces were all generally wax polished afterward, usually with a mixture of beeswax dissolved in turpentine, sometimes with various resins

added for durability and hardness. This wax polish gradually hardened, the wood tone shifted as it aged and a patina developed. Subsequent polishings maintained the finish, which mellowed as it acquired minute scratches, dents and a bit of grime.

A wax polish is compatible with nearly all old finishes, but how well a piece of

A wax-polished surface has a soft sheen unrivaled by any other finish. You can see the reflection of the buffing rag in the crotch-mahogany veneered surface of this drop-front desktop.

furniture responds to it will vary depending on how—and how well—a piece was finished originally and on the care (or lack of it) received by the piece since then. Pieces that originally were finished with care but not maintained well will respond to a polishing after being cleaned. Pieces that have either been restored in an insensitive manner (drenched in cheap varnish, for example) or finished poorly to begin with will probably need refinishing before they can benefit from a wax polish. I'll discuss below how I clean a piece of furniture, from the gentlest method to the most aggressive that I can recommend, and how I apply a wax polish, generally beginning with a color coat and then applying a harder clear wax over that. Properly done, a wax polish is the most beautiful surface treatment in the woodworker's repertoire (see the photo on p. 107).

Cleaning

Few pieces of old furniture are in perfect condition. If the dirt and grime obscure the grain, the piece needs a judicious cleaning before you polish with wax. I use a naphtha-soaked soft cloth to remove dirt, grime and built-up wax, right down to the old finish. (Naphtha is a petroleum-based solvent that's slightly more aggressive than mineral spirits; if you can't find it, mineral spirits will work fine.) I always start in an out-of-the-way place, preferring caution to speed. Finishes on old furniture are always completely cured, so there's very little chance of the naphtha dissolving them. I wipe the surface of the piece until the cloth comes off clean.

I've also removed old wax and dirt with Liberon's wood cleaner and wax remover (see Sources of Supply on p. 111). As long as it's not left on too long, this solvent won't harm the finish layer either. As with the naphtha, it's best to begin in an inconspicuous area to see how the finish responds to it.

If you find that you need to get more aggressive yet, substituting 0000 steel wool for the cloth will usually do the trick. Sprinkling the surface with rottenstone will increase the cutting action even more and will leave a very fine surface. Whenever I

use abrasives, I check the surface frequently (by wiping away the rottenstone and naphtha) to make sure I don't cut through the finish. I've never had a problem, though, probably because the finish layer on most antiques has had a hundred years or more to cure.

If a piece has areas of carving or intricate detail, I make a paste of rottenstone and naphtha and work it in with a soft toothbrush. Afterward, I remove all residue with pure naphtha on a soft, clean cloth. The naphtha sometimes leaves a slight film on the surface, but it will buff right off, and any traces will disappear when I apply wax.

Applying a color coat

After I have cleaned a piece and have given it at least 24 hours to dry, I apply what I call a "color coat" of wax (see the photo on the facing page). This color coat is only a preliminary step in preparation for a final coat of clear, harder wax. This first coat of tinted wax will hide minor dents and scratches, enhance the natural wood color and even out excessive differences in tone resulting from repairs or exposure to direct sunlight. A tinted wax will also fill any unfilled pores with darker particles than a clear or white wax will, resulting in a more natural-looking surface. When using a tinted wax, I select a shade darker than the wood tone I'm polishing because most of the wax is removed during the buffing.

I use two brands of colored waxes. One of them, Antiquax's Antique paste wax polish, produces a long-lasting film and doesn't fingerprint. I use the Brown wax most often because it works well over a broad range of wood tones, from light oak to dark mahogany. The other tinted wax I use is Liberon's Black Bison paste wax polish. This wax comes in ten colors (derived from natural earth pigment matter) as well as natural and clear (the natural is slightly amber while the clear is actually bleached—and quite clear). Georgian Mahogany covers quite a range of shades and is particularly useful on old English mahogany furniture, which I see a good deal of.

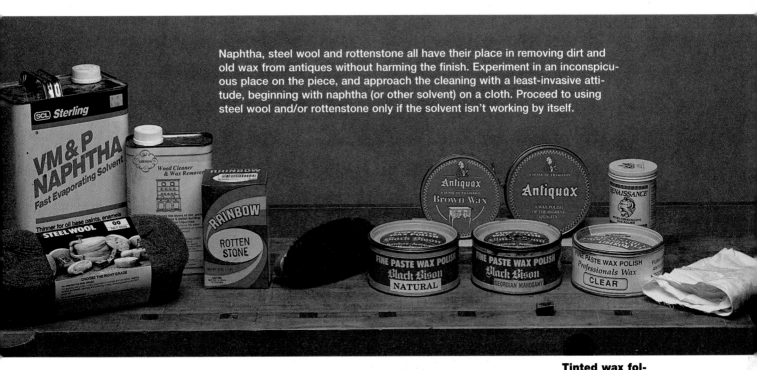

Naphtha, steel wool and rottenstone all have their place in removing dirt and old wax from antiques without harming the finish. Experiment in an inconspicuous place on the piece, and approach the cleaning with a least-invasive attitude, beginning with naphtha (or other solvent) on a cloth. Proceed to using steel wool and/or rottenstone only if the solvent isn't working by itself.

I apply the colored wax to one surface at a time, using an old cotton T-shirt and working in a circular motion, making sure that everything is covered. The colored waxes are relatively soft, so I apply them quite liberally, really working the wax into the surface. Antiquax recommends leaving its product on for two or three minutes only. Liberon suggests applying its product with steel wool and leaving it on for at least 20 minutes, giving the solvents time to evaporate before buffing. Optimally, leaving the wax on for four to eight hours allows thorough evaporation and will give you the highest sheen when buffed. I've left the color coat on overnight without a problem. It was considerably more difficult to rub out than if I'd followed the directions, but the resulting finish was harder.

I buff out the color coat with an old terry-cloth towel, rubbing vigorously and removing all but the finest layer of the colored wax. The harder I buff now, the deeper the luster of the finish. If a coat of wax does dry out before you get around to buffing it, no serious harm has been done. Applying a fresh coat will dissolve the hardened coat, and you can buff normally. With

experience, you'll determine how long to wait before buffing.

I save all my wax-impregnated rags in old cookie tins. A well-maintained antique—or one that you've brought back to life—often will need nothing more than a good rub with one of these cloths to bring back its luster.

For the color coat on the carved areas, I apply the wax with a soft toothbrush, and then I buff it off with a pure-bristle shoe-polishing brush or with one of Liberon's wax-buffing brushes.

A color coat may not be advisable on some pieces. Light woods—fruitwoods or maples, for example—look just fine as they are. On furniture made of these and other light woods, I use Antiquax Clear paste wax polish, which has a slight amber tone, but will not discolor these woods because buffing leaves such a thin layer of wax on the surface. Liberon's Neutral and Clear waxes are also good choices, as is Renaissance wax. Renaissance wax has a stiffer consistency than the other waxes I've discussed and can be difficult to apply evenly, but I circumvent this problem by applying it only to very small areas at a time and then buffing almost immediately. The polish it leaves is

Tinted wax followed by a clear wax can make a wood surface absolutely radiant. The author applies a color coat first, and then, after it's thoroughly dried, he applies a clear coat of harder wax. He buffs both the color coat and the clear coat with clean rags—preferably old linen napkins or tablecloth scraps—and buffs carvings and other relatively inaccessible areas with a shoebrush or one of Liberon's buffing brushes.

With the harder topcoating of clear wax, work in one small area at a time, and apply the wax polish in a tight circular motion. These polishes contain more wax and less solvent, so their drying times are considerably shorter.

beautiful, although not as resistant to water or alcohol as other waxes.

Applying a clear coat

It's essential that you use a clear coat of harder wax over the color coat because even if you've buffed the color coat thoroughly, there's a possibility of color transfer from the furniture to your clothing, especially on chair, table or desk edges. Just as important, though, is the protection the harder wax provides your furniture. I like to wait several weeks after applying the color coat before I apply the clear wax. It doesn't always work out that way, but that's the ideal. For this topcoat, I've been happy with Antiquax's Clear paste wax polish, Renaissance wax and Liberon's Clear Professional wax, which is higher in carnauba content (therefore harder) than the waxes in their Black Bison line.

I apply the clear coat to small sections, using an old cotton T-shirt, working in a circular motion and then buffing a little

before the wax has completely set—no more than ten minutes (see the photo above). Old linen napkins or tablecloth scraps seem to work better than anything else for buffing out the clear coat. They produce a superior shine, and the wax-impregnated rags just get better with time.

Problem cases

Occasionally I encounter a piece that does not respond well to a wax polishing, even after a thorough cleaning. Often such pieces will show dull or worn areas when polished with wax, or the wax may seem to sink in without effect. If the finish seems sound, I clean the piece again, removing the wax I've applied and any remaining dirt. Then I melt the wax I'm using (usually tinted) in a double boiler or glue pot, and I add about a tablespoonful of rottenstone (not pumice, which is much coarser) for each 8-oz. container of wax. After removing the wax from the heat source, I add about one-quarter cup

Carved, multi-faceted surfaces are ideal candidates for a wax polish. The character of the wax is such that it accentuates the texture of the carving rather than masks it, as is evident in the author's reproduction shown here.

of mineral spirits to thin the mixture and make it easier to work.

When the mixture has cooled, I apply a liberal amount to one surface at a time, buffing it in with a lambswool pad on an electric drill. I add more of the wax mixture whenever it starts getting thin on the surface. The rottenstone lightly abrades the wood, burnishing the surface and enhancing the effect of the wax. After applying the wax in this manner, I rub it into the surface by hand, using a clean, soft cloth and following the grain. I rub briskly and then buff the wax off when dry, again following the grain of the wood. I let the wax cure in a warm atmosphere for at least two weeks and then apply a clear coat of wax in the usual manner. The result is a surface that will remain beautiful for years. Do not use any spray polishes, oils or other maintenance products on the finished surface. The only care or maintenance your furniture needs at this point is a quick rub down with the rag you used for the clear wax polish, and less frequently (it will depend on wear), another coat of the clear wax.

SOURCES OF SUPPLY

LIBERON

Liberon Supplies
PO Box 86
Mendocino, CA 95460
(800) 245-5611

ANTIQUAX

Marshall Imports
PO Box 47
Crestline, OH 44827
(800) 992-1503

RENAISSANCE

Garrett Wade Co., Inc.
161 Ave. of the Americas
New York, NY 10013
(800) 221-2942

FOUR

Spray Finishing

Spray finishing requires a large investment in equipment, space, and practice. First, there's the spray gun and compressor or turbine. Those will run several hundred bucks or more for a high-quality system. Then there's a spray booth with an explosion-proof fan if you plan to spray oil-based finishes. And where to put them? Who has a shop with a hundred square feet or more of unused space? The alternative is to spray outside, where, unfortunately, you can't control the humidity (crucially important in spraying) and where random insects get very curious and consequently involved in your finish. Of course, if you're not paying attention, the wind can shift and overspray will coat the family car. Even then, learning how to spray well takes time—and the patience to sand out the inevitable mistakes.

So why on earth would any sane, amateur woodworker get entangled in spray finishing? Simply because all the trouble is worth it. Spray finishing offers levels of speed, quality, and versatility that hand-applied finishes can't begin to match. How much time do you spend brushing on a finish, letting it dry, rubbing it out, and then going through the process again and again? And how good do those finishes look? When done right, spraying can give you an almost perfect finish in a fraction of the time it would take to brush or wipe on. Like buying a tablesaw, the substantial investment you make up front will pay for itself every time you don't have to rip a board by hand.

The articles in this chapter will introduce you to spray finishing and get you well on the road to understanding and mastering the issues involved. Andy Charron gives you detailed reasons why spraying is the best way to finish. Chris Minick explains the advantages and disadvantages of the several major types of spray systems, including conventional high-pressure spraying, high-volume, low-pressure (HVLP) spraying, and airless spraying. Nick Yinger offers a simple way to turn a vacuum cleaner motor into a turbine. For about the same amount of time it takes to build an average workshop jig, you'll have an excellent HVLP setup for a fraction of the retail cost (though the spray gun is not included).

TAKING THE SPRAY-FINISH PLUNGE

by Andy Charron

My first shop was a one-car garage. What space I had was filled with tools that were absolutely necessary to make furniture. That left out a dust collector and a finishing room. As a result, getting dust-free finishes was frustrating. Brushing on shellac and varnish worked fine for small projects, but as I took on bigger jobs and built more pieces, I turned to wipe-on oils because they weren't as fussy to use. Eventually, I needed more durable finishes that didn't take long to apply.

Spraying has benifits over other methods of finishing

Spraying gets finish in nooks and crannies. One reason Andy Charron switched to spraying is that it gets finish where other applicators won't. Here, he sprays water-based sealer on the latticework of a poplar headboard.

1) Spray finishes are forgiving. Because a sprayed finish is built up in thin layers, small scratches and marks stay better hidden under a sprayed translucent finish than under an oil finish. Surface preparation is still important, though. This is especially true when spraying paints or opaque stains.

2) Spray finishes are fast. You can spray 30 stools or 1,000 small wooden blocks in an hour. And because the sprayer breaks the finish into small particles, each coat dries in a hurry. Many varnishes, water-based products and sprayed lacquers will dry to the touch in minutes. Some of them can be sanded and re-coated in a few hours. Dust has a short time (while the coat is tacky) to settle on the work, which reduces the need for sanding between coats.

3) Spray finishes are versatile. Basically, any finish that can be applied by brush or by rag can be sprayed. If you use an explosion-proof booth, you can spray shellac, lacquer and other solvent-based materials. If you don't have a booth, you can still spray water-based finishes. With some spray sys-

A spray system was the answer. Spraying on finish is fast and easy. You can get into places where brushes and rags are useless (see the photo on the facing page). Spray finishes look superb, too. The coating is more uniform and the finishes between pieces is more consistent. But once I was committed to changing to spray finishes, I knew I had some research to do.

Spray systems and finishes are better now

The variety of spray systems has increased dramatically over the last 10 years. Manufacturers have introduced small, inexpensive units that are ideal for hobbyists and small shops. Also, there have been many improvements in high-volume, low-pressure (HVLP) spray systems, particularly in terms of transfer efficiency. The price of an entry-level HVLP spray system is around $200, and there's a wide variety of systems in the $200 to $500 range. These spray systems aren't much more expensive than many power tools.

Waterborne finishes have improved as well, and as a result, the need for dangerous, solvent-based finishes has decreased. Water-based finishes are nonflammable, which means that you no longer need a spray booth to get started. Having a clean spray area, a respirator and good ventilation (I use an exhaust fan) will suffice. And a spray system won't leave you with a pile of oily rags that can catch on fire.

Brush-on and wipe-on finishes are slow and exacting

In my furniture business, I brushed on varnishes for only a short time. Varnish was just too slow to brush and too slow to dry. And I needed excellent lighting to brush, sand and rub out the varnish.

Spray guns increase production. Charron compares the number of clock frames he sprayed (left) vs. those painted with a brush.

tems, you can apply water-based contact cement, which works great for laminate work.

4) Spray finishes can be precisely controlled. Spray-gun adjustments combined with proper spray techniques give you good control over how and where the finish is applied. A brush transfers nearly 100% of the finish to the work, but you have to be diligent at keeping the coat even and at the right thickness.

Even though the transfer efficiency of a spray gun is lower than a brush (between 65% and 85%), you can adjust air pressure, fan size and fluid flow to ensure light, even coats. Because the atomized material flows together uniformly, there are no brush or lap marks.

5) Spray finishes are relatively easy to apply. Spray finishing is fairly basic. You can learn how to spray a simple case or frame in less time than it takes to master brushing or wiping on a finish. With a bit of practice, you can spray stains and dyes to get uniform coverage and consistent color depth. After some more practice, you can use tinted clear finishes to do spe-

cial techniques, such as shading or sunbursts. Because spraying allows a greater range of finishes, your projects will look more professional.

6) Spray finishes are consistent in quality. Without a doubt, the best reason for investing in a spray system is the overall higher quality of finish that you can achieve.

A spray-on finish is far superior to brush-on or wipe-on finishes. The problems caused by brushing, such as runs, drips and air bubbles, are reduced with spray equipment. And brush marks are gone. You can spray an entire piece, no matter what its size or shape, with light, even coats of finish.

I did stick with wipe-on oils for a while. Oil didn't require any special equipment, and I could oil in less-than-ideal conditions. I wasn't building up a thick surface film (like a varnish), so I worried less about dust and lint getting trapped in the film. Oil finishes soon became a key in my marketing strategy, too. Most of my customers liked the phrase, "authentic, hand-rubbed finishes."

Oil finishing does have drawbacks. The protection offered by an oil finish is minimal, and an oil finish needs more maintenance than other topcoats. Surface imperfections, like scratches, stand out more than they would under a film finish. And oil finishes are time- and labor-consuming. Depending on the temperature and humidity, an oil finish can take several days to apply. It also involves a great deal of work. It's hard to get thrilled about rubbing out multiple coats of oil on 400 wooden clock frames.

Clean finish, clean gun—To get blemish-free finishes, the author filters the finish before he sprays, and he cleans the gun afterward. He often tints his paint basecoats with pigment, so the topcoat covers better.

Any spraying disadvantage can be overcome

As attractive as spray finishing is (see the box on pp. 114-115), it does have a few weaknesses. Setting up a safe, efficient system takes up shop space and costs money. Besides a gun, you will need a source of air (either a turbine or compressor), hoses, filters and connectors. Because spraying releases finish mist into the air, you will also need a spray area that has fresh-air circulation. If you spray solvent-based finishes, you'll need to check with your local building inspector before you set up a booth. But if you spray water-based products exclusively, you won't need explosion-proof fans and fixtures.

Unlike most brush-on and wipe-on finishes, spray finishes must be filtered and then thinned to the correct viscosity (see the photo at left). Not thinning enough can lead to lumpy finishes and "orange peel." Using too much thinner creates problems, too, like drips and sags on vertical surfaces. And it will take longer to build to the right film thickness. The result is you won't be able to get nice, glossy clear coats, and paints won't hide the underlying surface or provide good color depth. Too much thinner also lengthens the drying time, so dust becomes a problem.

Finally, keeping your spray gun clean is critical. Although cleaning does involve some effort and time, it doesn't take any longer to clean a spray gun than it does a brush.

Ultimately, spraying reduces finishing costs

Although some of the finish does get wasted through overspray, you can still lower your material costs. I've had to reject far fewer pieces that I've sprayed than those that were finished by brush or rag. And spraying saves labor costs. In the first month, I more than offset the initial expense of the equipment (about $800). Now my business couldn't survive without a spray system.

WHICH SPRAY SYSTEM IS RIGHT FOR YOU?

by Chris A. Minick

Mention the names Delta, General or Powermatic to a bunch of cabinetmakers, and everyone in the group will know you're talking about woodworking machinery. Mention DeVilbiss, Mattson or Sharpe to the same crowd, and you'll likely get some blank stares. Those companies are just three out of dozens that make spray-finishing equipment. Chances are, though, many woodworkers just don't know as much about choosing a spray system as they do about buying a tablesaw. Considering that a high-quality spray system costs as much as a decent tablesaw ($700 or more), it pays to be well-informed before you buy.

Andy Charron explains why he switched to spray finishing in his shop (see pp. 114-116). I'll present some equipment options—high-pressure spray guns (see the photo at right), high-volume, low-pressure (HVLP) systems and airless spray guns. But first, it would be helpful to know a little about spray-gun anatomy.

How a spray gun works

The basic principle behind a spray gun is relatively straightforward. A stream of liquid finish is forced into an airstream, which breaks the liquid into tiny droplets (atomization) and carries them to the target surface. It sounds simple, but in reality, a collection of precision parts must work in concert to pull the whole thing off.

In a standard high-pressure system, air flows from the compressor hose through a series of valves and baffles in the body of the gun and out through an air cap. The valves and baffles control the maximum atomiza-

Use a booth when spraying solvent-based finishes, such as nitrocellulose lacquer. Here, the author uses a Binks high-pressure spray gun, which has a 1-gal. paint pot. These guns produce excellent results but lots of overspray.

An air compressor can power a high-pressure or conversion-air HVLP spray system. With either type, you'll need an oil and water filter separator, a regulator, an air hose and couplers. Choices for guns (from left): conventional touch-up, external and internal mix, two HVLP units and conversion-air touch-up.

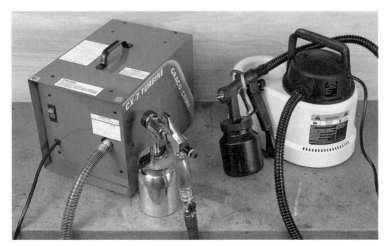

Turbine-driven HVLP systems are compact, but the hoses are cumbersome. Both the two-stage Graco/Croix unit (left) and the Wagner single-stage model spray efficiently and are portable.

tion pressure at the air cap. The volume of air used by the gun as well as the spray pattern is governed by the size and placement of the holes in the air cap (see the drawing on p. 120).

A standard air cap for furniture finishing produces a tapered (fan) pattern 9 to 11 in. long. Typically, the gun uses about 8 cubic feet per minute (cfm) of air at 50 psi.

Pulling the trigger extracts the needle from the fluid tip, which opens the orifice and allows the finish to enter the airstream. The size of the orifice and the viscosity of the finish control the amount of material sprayed. I've found that a 1mm orifice is ideal for finishing furniture. The fluid tips and needles are sold in matched sets (fluid setups). Most spray-system manufacturers have technical-service departments that will help you choose the right one.

Once the finish exits the tip, high-pressure air from the air cap blasts the stream into tiny droplets. The droplets can range from about 15 microns in dia. to 70 microns or more. The size depends on the fluid viscosity and on the equipment. Once the atomized finish is deposited, it flows together to form a smooth film. Generally, the smaller the droplets, the better the finish.

Gun composition affects the kind of finishes you can spray

A gun that has an aluminum cup and fluid passages is compatible with hydrocarbon-solvent-based finishes like nitrocellulose lacquer and oil-based varnish. But within a matter of hours, the same gun will be corroded beyond repair if it is used to apply a finish that contains a chlorinated solvent, such as methylene chloride (which is the main ingredient in many paint strippers). Even nonflammable solvent cleaner will corrode aluminum parts. Similarly, the alkaline portion of waterborne finishes can damage bare aluminum parts if the gun is not cleaned immediately after use.

As a corrosion-fighting alternative to aluminum, some low-cost units combine plastic cups and dip tubes with brass fluid-handling parts. But brass wears quickly, particularly if the gun is used to spray pigmented finishes like paint. The pigments act like the abrasives used in sandblasters.

Mild-steel components (especially fluid tips and needles) are also common in inexpensive spray guns. Though steel is compatible with most finishes, it has a nasty tendency to rust. One solution is to buy a gun that has a stainless-steel cup and fluid-handling parts, but that type is pricey. Those guns make sense for industrial users, but they are overkill for small shops. As an alternative, some spray guns come with stainless-steel fluid passages and a Teflon-lined aluminum cup. The Teflon lining protects the cup from corrosion and makes for easy cleanup.

High-pressure spray equipment

Early in this century, high-pressure spray equipment was developed in response to the automotive industry's need for high-speed finishing. Spray components have changed little since that time (see the top photo on the facing page). A full system consists of three main parts: a compressor (with attendant hoses, tank and pressure regulator), an oil and water separation device, and a spray gun.

The air compressor is the heart of the spray system; both the horsepower rating and tank size affect spray performance.

A 3-hp compressor with an air output of 10 cfm and a 20-gal. air tank is really the minimum size.

When air is compressed, water vapor in the air condenses to a liquid. If not removed, the water that passes through the spray gun will cause all kinds of finishing problems. So an oil and water separator is a critical part of any compressor-driven spray system. The separator also removes residual oil that's used for lubrication of the compressor.

Internal mix or external mix

High-pressure spray guns are available in two types: internal and external mix (see the photo and drawings on p. 120). The mix designation is based on where the airstream is introduced into the fluid stream.

Most internal-mix guns (air and fluid are mixed inside the air cap) produce a coarsely atomized spray. Although this spray is unsuitable for applying lacquers or other fast-drying finishes, it is ideal for applying thick, difficult-to-spray materials, like adhe-

sives and pore fillers. Internal-mix guns consume modest amounts of air and can be powered with a 1-hp or 2-hp compressor. But they are limited to spraying slow-drying varnishes and paints.

By contrast, external-mix guns (air and fluid are mixed outside the air cap) are versatile. They're the most common spray guns used in woodworking shops. Hundreds of fluid tip/needle/air-cap combinations are available to allow the spraying of virtually any liquid at almost any pressure. External-mix guns can be fed from a 1-quart siphon cup attached to the gun or pumped from a 1-gal. remote pressure pot when greater quantities are needed.

External-mix spray guns have two drawbacks. They use lots of air, so they require at least a 3-hp (4 hp or 5 hp is preferable) compressor. And they aren't very efficient at putting the finish on the work. Only about 35% of the finish actually lands on the target; the rest ends up as overspray. High-pressure spray guns only make sense in a shop that has a good spray booth.

More finish ends up on your project with HVLP

High-volume, low-pressure (HVLP) spray equipment has been around a while. In the late 1950s, I painted models and birdhouses with an HVLP painting attachment that came with my mother's canister vacuum cleaner. HVLP equipment is more sophisticated now, but the underlying concept remains unchanged. To atomize the finish, HVLP systems use high volumes of air rather than high pressure. Unlike conventional spray guns, HVLP guns produce a soft spray pattern. The benefits are increased transfer efficiency, low overspray and almost no bounce-back. Simply put, HVLP spray

Airless spray systems work well with latex paint and most varnishes, but they don't apply other finishes well. If not the right viscosity, the finish will be poorly atomized and leave a coarse, blotchy surface.

guns put more finish on the project and less on everything else in the shop and in the environment.

Spray-equipment manufacturers have taken two very different approaches to HVLP. Some have developed turbine-driven systems and others have developed conversion-air HVLP systems, which are driven by a standard air compressor.

Turbine-driven HVLP spray systems are portable

Turbine HVLP systems use a fan (like those used in vacuum cleaners) to generate from 45 cfm to 110 cfm of air at pressures between 2 psi and 7 psi. You can buy turbines in three power levels: one, two or three stage. The bottom photo on p. 118 shows a two-stage turbine and a single-stage unit. Each stage, or fan section, in the turbine adds approximately 40 cfm and 2 psi of air output.

Unlike a compressor, a turbine blows out a continuous stream of warm, dry air at a constant pressure. This eliminates the need for pressure regulators and air dryers (separators). But warm air can be a problem. The metal handles of some spray guns can get uncomfortably hot.

Also, dried drops of finish tend to accumulate on the fluid tip; eventually, the finish glob breaks free and deposits itself on the freshly sprayed surface. On the positive side, turbine systems are compact, store easily and operate on 110v current.

The more stages a turbine has, the wider the viscosity range of the spray finish. When I sprayed with a one-stage turbine (a Wagner Finecoater), I had to thin the finish to get proper atomization. Thinning is the kiss of death for some waterborne finishes. When I sprayed the same finish with a two-stage turbine (a Graco/Croix CX-7) there was sufficient power to spray without thinning. I didn't try a three-stage turbine. Designed for multiple guns and high production, they're a bit pricey for me (more than $1,000).

Conversion-air HVLP spray systems are versatile

Conversion-air HVLP systems convert compressed air (under high pressure) to a high volume of air (at low pressure) by passing it through baffles and expansion chambers in the gun body. A decent gun costs $250 or more. Conversion-air guns have the reputation of being air hogs. But the latest conversion-air spray guns will operate off most 3- or 4-hp compressors. If your shop already has a compressor, it may power a conversion-air HVLP gun.

A big advantage that conversion-air systems have over turbines is that the atomization pressure at the air cap can be adjusted (between 2 psi and 10 psi with most guns) to accommodate a wide range of coating viscosities. I compared the two types of HVLP systems side by side (see the box on the facing page). The conversion-air system consistently produced a finer atomized finish, a higher delivery rate and a noticeable decrease in overspray.

Conversion-air spray guns work best when connected to $^3/_8$-in. air hoses. The quick-connect fittings on the hose and the spray gun must be matched (connectors are

Air caps

Guns can be internal mix (left) or external mix (right). The spring, retaining ring and baffle have been removed in the external-mix gun.

Internal-mix air cap
Air and finish are mixed inside cap.

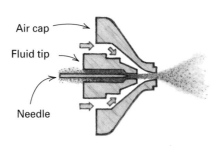

Air cap

Fluid tip

Needle

External-mix air cap
Air atomizes finish and shapes spray pattern outside cap.

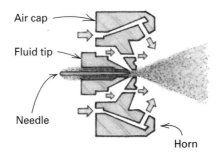

Air cap

Fluid tip

Needle

Horn

available at most auto-paint and compressor repair shops). Use a ³/₈-in. coupler; a ¹/₄-in. coupler will negate the advantage of the larger hose.

Airless spray systems

Airless systems usually are associated with house painting rather than furniture finishing. But airless spray systems are common in large furniture factories. These commercial units operate at pressures approaching 4,000 psi. However, high pressure, high delivery and high efficiency come with a high price tag—upward of $1,500 for an entry-level unit.

Consumer-sized airless units (see the photo on p. 119) still have a place in the shop. I like them for applying latex paint and oil-based varnish on certain projects. The motor size of an airless gun determines its price and its versatility. A 110-watt gun is powerful enough to spray unthinned latex paint. But with a 45-watt unit, the paint has to be thinned significantly. A motor rating of 85 watts or more usually is adequate for spraying furniture.

Unfortunately, airless spray guns produce a coarse spray pattern. So only slow-drying paints and varnishes should be applied with them. Lacquers, including waterborne varieties, tend to dry before the droplets flow together. The result is orange peel.

Even with these limitations, however, an airless spray system can help get you started spray finishing—and for a reasonable price (around $200). One of the best things about an airless spray unit is that it doesn't use a cumbersome air hose. It just needs an extension cord.

The choice is yours

If you're considering a spray system for your shop, take a good, hard look at conversion-air HVLP spray systems. As a bonus, you'll have an air compressor to do other things in the shop.

Evaluating spray patterns

I couldn't resist comparing the performance of the spray systems in this article. I used a gloss, water-based lacquer (tinted black) in each spray gun. This is a demanding test when you consider I didn't adjust the viscosity. Spray patterns reveal where atomization was poor (large spots on borders) and where fan adjustments were limited (wide dispersion band). In general, high-pressure and conversion-air HVLP systems delivered fine atomization and more uniform spray patterns. Turbine HVLP and airless systems produced coarser spray patterns.

Coarse and splotchy

Fine and uniform

Conventional high-pressure spray (from Cal-Hank touch-up gun).

Conversion-air HVLP spray (from DeVilbiss touch-up gun).

Turbine HVLP spray (from Graco/Croix gun). Finish was thicker than recommended viscosity.

Airless spray (from Wagner gun). Finish was thinner than recommended viscosity.

BUILD AN HVLP TURBINE WITH A VACUUM MOTOR

by Nick Yinger

Shop-built spray unit—A high-volume, low-pressure unit like this one that the author built is ideal for on-site work or in the shop.

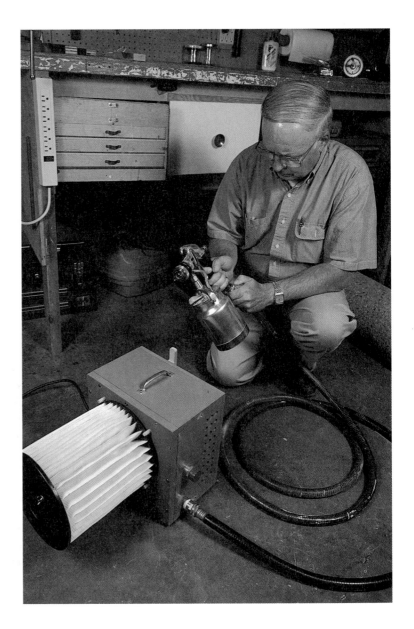

For years, I did my spray finishing with a conventional compressor-driven setup. I was never entirely satisfied with the arrangement, and I recently built my own high-volume, low-pressure (HVLP) unit, as shown in the photo on the facing page, to replace it.

What bugs me about conventional spraying? For starters: finishing the inside of a case with a swirling cloud of overspray billowing back in my face. I can't see what I'm doing, and I wind up ingesting a big dose of chemicals no matter what kind of mask I wear. Even when I'm spraying water-based finishes, which are inherently safer, I find overspray annoying.

Although they're neither toxic nor flammable, water-based finishes are expensive, so it makes even less sense to blast these precious fluids all over the booth with air compressed to 50 pounds per sq. in. (psi). HVLP spraying looked like the answer to these problems. This method promised to transfer 70% to 80% of the material from the gun to the object compared with 20% to 30% with a conventional setup. To accommodate a stream of warm, dry, low-velocity air, HVLP guns have large hoses and air passages. They use copious amounts of air—as much as 30 cu. ft. per minute (cfm) but at only 5 psi. (For pros and cons of HVLP, see the box on p. 124)

I had a 3-hp compressor, so it seemed a simple matter to install a large, low-pressure regulator to feed 5-psi air to the gun. But there was a catch. A 3-hp piston compressor won't pump 30 cfm continuously at any pressure. The rule of thumb is 1 hp per 4 cfm of air, and we're talking about large, healthy, industrial horses not puny, underfed, home-improvement horses. Because 8- to 10-hp compressors are expensive and connecting my small compressor to a tank the size of a submarine seemed impractical, I decided I'd investigate the turbine compressors sold with HVLP guns.

I borrowed an HVLP unit from a friend and used it to finish some bathroom cabinets. It performed beautifully: almost no overspray, good atomization and good fluid and pattern control. My only criticisms were that the hose seemed cumbersome, and the handle of the gun became uncomfortably hot.

As I used the HVLP unit, I couldn't help thinking that if it acts like a vacuum cleaner, sounds like a vacuum cleaner, it must be a vacuum cleaner. I peeked inside. Sure enough—a two-stage vacuum cleaner turbine with an 8-amp motor! Soon thereafter, I set out to build my own HVLP turbine compressor.

Build your own HVLP unit

An HVLP machine is a centrifugal turbine compressor contained in a box with an inlet to bring air into the turbine and a plenum or outlet chamber to capture the compressed air discharged by the turbine and route it to your sprayer hose (see the drawing on p. 127). The turbines used in large vacuum cleaners are integral with their electric motors and are referred to as vacuum motors.

First buy a vacuum motor

Go to an industrial supply company, or get their catalog. I bought mine at Grainger (contact their marketing department at 333 Nightsbridge Parkway, Lincolnshire, Ill. 60069; 800-473-3473 for the nearest location); their catalog lists 45 vacuum motors, ranging from $40 to more than $280. You'll find a wide selection of features, such as bearing type, motor voltage, number of compressor stages and motor amperage. Most important for this application is bypass, not flow-through motor cooling. This means the motor is cooled by a separate fan. With this design, the motor won't overheat if the vacuum inlet or outlet is obstructed.

Single-stage compressors move large volumes of air but produce the lowest pressure. Two- and three-stage units supply higher pressure air at some sacrifice in volume but typically have more powerful motors and, hence, better overall performance. I chose a two-stage turbine with a 13-amp motor rated at 116 cfm that costs $163, an Ametek model #115962. I could have purchased a less powerful unit, but I wanted to be able

Conventional spraying vs. HVLP

by Dave Hughes

It's 8 a.m., and you've just entered your shop, coffee in hand. Standing before you is your latest project, nearly completed. It just needs to be lacquered. You take a deep breath, fill your spray gun, crank up the compressor, put on a particle mask and go for it. Fifteen minutes later, the atmosphere in your shop resembles that of Venus: Every tool is covered with a fine white dust, the shop's out of commission for the rest of the morning and you've got a serious headache. Sound familiar? If, like most of us, you've tried to do finishing with conventional spray equipment in a small space, it probably does. There's an alternative. It's high-volume, low-pressure (HVLP).

By now, most professional finishers have an HVLP unit in their arsenal of tools and increasingly, the units are finding favor with folks who do only occasional finishing. One big reason is that HVLP units have far higher transfer efficiency than conventional spray units. This means, simply, that most of the stuff you're spraying goes where you want it to go. A painter friend of mine did his own little test when HVLP first hit the market. He painted one cabinet with a traditional, compressor-driven gun and an identical cabinet with an HVLP unit. When he was done, there was three times as much paint left in the HVLP cup. Where was the paint missing from the conventional gun? All over.

Aside from transfer efficiency, HVLP offers a string of clear benefits over conventional setups:

• They are compact, lightweight, self-contained, easy to set up and clean.

• The guns have a wide variety of spray-pattern settings for finishing intricate shapes as well as broad, flat surfaces.

• The low-pressure air supply is adjustable and so creates far less bounce-back of material from inside corners.

• The dry, heated air helps materials flow on smoothly, level out nicely and set up quickly. It also helps avoid blushing on cold, damp days.

• Your shop is not rendered useless for hours. (But open a window anyway.)

Drawbacks? There are a few:

• HVLP units are not really a high-production tool but are more suited for small-to medium-sized projects.

• Standard models have a rather cumbersome air hose all the way to the gun, limiting wrist mobility somewhat.

• As with any quart-gun arrangement, you can't spray upside down, and you're constantly, it seems, filling it up. (Higher priced models offer a 1- or 2-gal. pot that stands on the floor for less-restricted gun movement and less-frequent fill ups.)

• And there's that whining motor—it reminds me of a car wash vacuum.

HVLP is a definite advance for the small-shop woodworker or finisher who wants professional results. With prices starting under $500 and savings from high transfer efficiency, they're a good investment. From the money you save, stake yourself fifty bucks for a decent charcoal respirator and a pair of earplugs.

High volume, low pressure (HVLP) in a small package. At 15 in. sq. and 18 lbs., the shop-built turbine-powered HVLP spray unit in the photo below is a fraction of the size and weight of the standard medium-sized compressed air setup in the photo at left.

to operate two spray guns on occasion, and anyway, I like overbuilt machinery. For a one-gun setup, you might try the Ametek 115757-P, which costs $63. For the rest of the parts in my HVLP unit, including the hose but not the gun, I spent less than $70.

Make a cradle for the motor

These motors are designed to be mounted by clamping the turbine housing between two bulkheads using foam gaskets. Make the rear bulkhead first. Cut it to size, bandsaw the circular hole and then chamfer the back side of the hole. The chamfer will ease the flow of motor-cooling air away from the motor housing. Cut the positioning ring to size, and rough out the hole with the jigsaw, leaving it slightly undersized. I made a Masonite routing template to exact size by cutting the hole with a fly cutter on the drill press. Use the routing template to finish the hole in the positioning ring.

Cut the housing sides, top and bottom to size, and make the dado for the rear bulkhead in each of them. Then drill the cooling outlet holes in the side pieces. Assemble the housing with the rear bulkhead in place, and when the glue has set, drop in the positioning ring, and glue it in place. I used screwed butt joints for the housing pieces and relied on the bulkhead to stiffen the box.

Mounting electricals—Switch, cord and circuit breaker are mounted in the back panel. Holes in the side of the back chamber are for motor-cooling air. A wooden cleat holds the wound cord.

Gasket and sealant

The turbine is held in the circular rabbet created by the bulkhead and positioning ring and is isolated from the wood by silicone rubber sealant. To hold the turbine centered in the rabbet while the silicone sets, cut three 2-in.-long pieces of $1/8$-in.-inside-dia. (ID) soft rubber tubing that compresses to about $1/16$ in. under moderate pressure. (This surgical tubing, with a wall thickness of $1/32$ in., is available in hobby shops and medical supply houses.) Lay the housing on its back, and put a generous bead of silicone in the rabbet. Lay the three pieces of tubing across the rabbet at 12 o'clock, 4 o'clock and 8 o'clock, and push the turbine down into the wet silicone. If you want the turbine to be easily removable later, spray the rim with an anti-stick cooking spray such as PAM before setting it into the silicone. Let the silicone set, and trim off the squeeze-out and tubing ends later.

Next rout the gasket grooves around the front edge of the housing, and press lengths of $3/16$-in.-ID soft rubber tubing into them. Make the front and back covers, and apply the rings of $1/2$-in.- by $1/2$-in. adhesive-backed weatherstrip, as shown in the top photo on p. 125, and then screw on the front and back.

Holes in the box

I tried various locations for the outlet holes and found no detectable differences. But I did get better output when I installed a fairing made from a strip of plastic laminate, which makes the outlet chamber roughly cylindrical (see the top photo on p. 125). Drill one or two 1-in. outlet holes in the housing, and screw $3/4$-in. pipe thread close nipples into them. Attach adapters to the nipples to provide $3/4$-in. male hose threads.

I attached a large shop-vacuum air filter to the front cover. Four short dowels hold the base of the filter in place, and a bracket pulls it tight against the cover. The bracket consists of two threaded rods screwed into the front cover joined by a hardwood crosspiece with a bolt through its center. A washer and wing nut secure the closed end of the filter against the crosspiece. You could also try using a large automotive filter. In that case, a Masonite or plywood disc secured by a similar bracket could hold the filter against the front cover.

Electricals

Mount the electrical parts: a heavy-duty switch, a circuit breaker with the appropriate rating for your motor, and the supply cord through the back cover, as shown in the bottom photo on p. 125. Then add rubber feet, a carrying handle and a cord-storage device.

Nice hose

I tried three different types of hose. All were $3/4$ in. ID and can be equipped with ordinary garden hose threaded fittings or quick-connect couplers. The most flexible was the lightweight, corrugated type provided with most factory-built HVLP sprayers, but its rough inner surface doesn't deliver as much air as smoother types. Plastic garden hose is cheap, smooth inside and flexible when warm, but in use, the heated air causes the hose to become too soft and to kink easily. My favorite is Shields Vac extra heavy duty/FDA hose available from marine distributors. It is made of a soft flexible vinyl molded around a hard vinyl helix. It's recommended by the manufacturer for use in boat plumbing below the water line, which means it will withstand a lot of heat as well as mechanical and chemical abuse.

Gun control

You can't just hook up your old gun to your HVLP turbine. HVLP guns are designed to enable them to atomize fluids with low-pressure air. List prices for these guns start at around $250. Of the HVLP guns I've tried, my favorite is a DeVilbiss (contact DeVilbiss at 1724 Indian Wood Circle, Suite F, Maumee, Ohio 43537; 800-338-4448 for a local supplier). The current model most like mine is their JGHV 5285 that lists for $365. It has stainless-steel fluid passages and a stainless-steel needle, so water-based finishes won't cause corrosion. And much to the relief of my palms, the handle is a nylon composite that doesn't get hot in use.

Shopmade HVLP unit

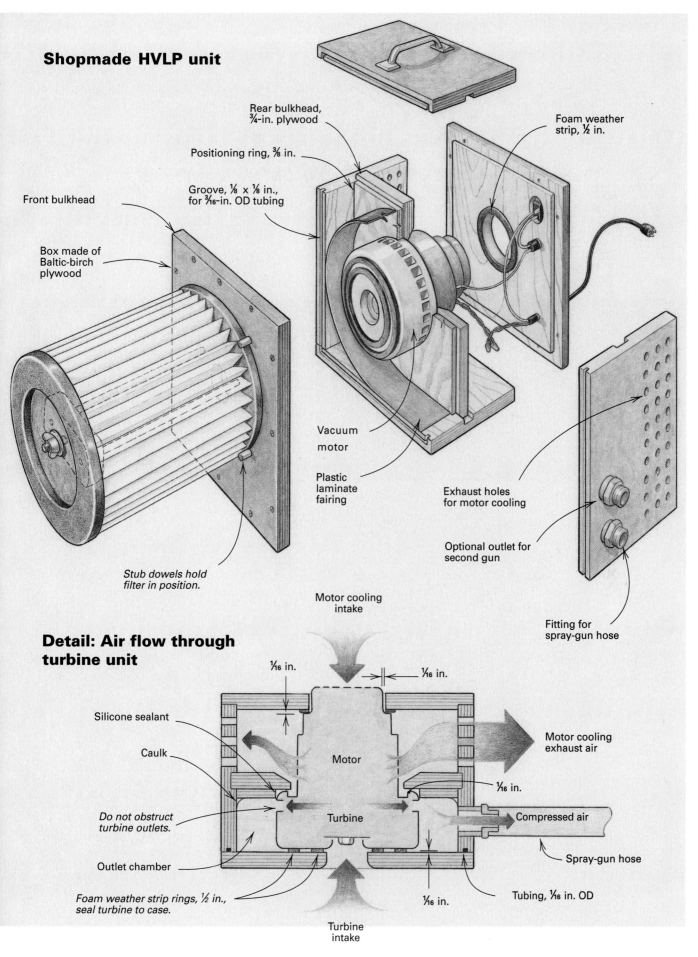

Rear bulkhead, ¾-in. plywood

Positioning ring, ⅜ in.

Groove, ⅛ x ⅛ in., for ³⁄₁₆-in. OD tubing

Front bulkhead

Box made of Baltic-birch plywood

Foam weather strip, ½ in.

Vacuum motor

Plastic laminate fairing

Exhaust holes for motor cooling

Optional outlet for second gun

Stub dowels hold filter in position.

Fitting for spray-gun hose

Detail: Air flow through turbine unit

Motor cooling intake

Silicone sealant

Caulk

¹⁄₁₆ in.

¹⁄₁₆ in.

Motor

Motor cooling exhaust air

¹⁄₁₆ in.

Turbine

Compressed air

Do not obstruct turbine outlets.

Outlet chamber

Foam weather strip rings, ½ in., seal turbine to case.

¹⁄₁₆ in.

¹⁄₁₆ in.

Tubing, ¹⁄₁₆ in. OD

Spray-gun hose

Turbine intake

TECHNIQUES FOR BLEMISH-FREE SPRAYING

by Andy Charron

Quite a few woodworkers I know are unenthusiastic, even fearful, about spray finishing. They believe the equipment is too mysterious, too costly and too hard to master. In fact, just the opposite is true. There are many simple-to-operate, reasonably priced spray systems out there. It took me less time to become proficient with a spray gun than it did to master a router. Best of all, the finish from a gun is often so smooth that I don't have to rub it out. Following sound spraying principles and knowing how to use the equipment helps me produce virtually flawless finishes.

Where to spray

The best place to spray is in a booth where a powerful exhaust removes overspray and dust from the air. If you're spraying solvent-borne finishes, you really have no other choice than to use an explosion-proof spray booth. But they're costly. You don't need explosion-proof equipment to spray waterborne finishes, and they're getting better and better. You only need a place that is well-ventilated and clean. If you have the floor space, you can build a spray room that has an exhaust fan and intake filters to ensure a steady supply of clean, fresh air. No matter where you plan to spray, check with your local building officials first.

Careful preparation is essential

How you prepare the surface is just as important as how you spray the finish. Sand the entire piece thoroughly (see the photo on p. 130). For stained work, I usually raise the grain with a damp cloth, let the surface dry and sand with 220-grit before I spray. For waterborne finishes and dyes, I sand to 180-grit and spray a light coat of dye stain or finish. This raises the grain and stiffens the fibers, making them easier to sand with 220-grit.

Spraying paint or pigmented lacquers is more involved. Opaque finishes highlight tiny imperfections. They often require at least two rounds of filling, sanding and priming before the wood is ready to be sprayed.

Thin the finish
to a sprayable consistency

Life would be easier if you could always pour finish straight from the can into a spray pot and begin applying it. But occasionally, you'll have to thin it. Which thin-

Spraying takes a bit of practice. Surface preparation, finish consistency and technique all are important.

ner you use and how much you add will depend on the material you're applying, the spray system you're using and what the piece will be used for. Some manufacturers do a lousy job of providing thinning information. If the appropriate thinner is listed on the label, use it. Because some cans of finish say that the contents don't need to be thinned, they don't list a thinner. If this is the case, you generally can thin the finish with the solvent that's recommended for cleanup.

Finding the correct viscosity

The viscosity of a finish is a measurement of its resistance to flow. Thinning a finish lowers the viscosity, which allows it to be broken into smaller particles (or atomized) more easily by the spray gun. The finer the atomization, the smoother the appearance.

Thinners can eliminate common spray problems (see the box on pp. 132-133) like orange peel, but if used improperly, thinners actually cause problems. Waterborne finishes are especially sensitive to thinning. Overthinning can prevent the finish from forming a clear, hard film.

Some spray-gun manufacturers recommend finish viscosity for a particular needle/tip combination. This information may be given as a ratio or a percentage of thinner and finish. The viscosity also may be given as the number of seconds it takes to empty a certain size viscosity cup.

Viscosity cups have small holes in the bottom, which let liquid drain through (see the photo on the facing page). Appropriately sized cups are available from most spray-system makers.

Room conditions are a factor

Temperature and humidity dramatically affect how much thinner to use in a finish and how it will spray. Low temperature and high humidity are not especially conducive to spraying. Even if you follow all the labels exactly, you may have to adjust the amount of thinner you add. You can keep records of how much thinner you need for different conditions. After a while, you'll get a feel for this.

Straining the finish and filling the pot

Your finish and your equipment should be as clean as possible because a speck of dirt or dried finish could ruin the job. To remove impurities, pour the finish through a strainer or filter (available at paint-supply stores). As an added precaution, you can install a filter on the end of the dip tube that draws finish from the pot, or put an in-line filter near the gun. To keep the air that comes from the compressor dry and clean, I run the line through a canister-type separator, which filters out water, oil and dirt before they get in the hose supplying air to the gun.

Sprayed finishes are only as good as the surface below. The author primed this bookcase and now sands it with 220-grit paper in preparation for spraying on a tinted waterborne lacquer topcoat.

Selecting suitable fluid tips and air caps

The fluid tip in a spray gun controls the amount of finish that gets deposited on a surface. In general, lighter finishes require a small tip. Thicker materials (or those with a higher percentage of solids) require larger fluid tips. The air cap in a spray gun controls the velocity of the air, which governs how finely the fluid is atomized. Air caps with smaller holes cause the air to leave the gun at a higher velocity, thus producing finer atomization. Air caps are matched with fluid tips to give optimum performance.

Most guns come equipped with a standard setup appropriate for several finishes. The setup includes a fluid tip that's about 0.050 in. dia. and a corresponding air cap. The standard setup will produce acceptable results with most finishes, but sometimes it's worth trying other combinations of fluid tips and air caps.

In a turbine-driven high-volume, low-pressure (HVLP) system, the amount of air feeding the gun is constant, so adjustments to the air pressure can only be done by changing air caps. If you are using a waterborne finish with a turbine and a bleeder-type (constant air flow) gun, make sure that the nozzle stays clean. These guns are prone to blobs of finish drying on the air cap and then blemishing the work.

Adjusting the gun

Spray guns come with adjustments for air and fluid. The type of finish being sprayed, the size of the object to be coated and the speed of application all play a role in deciding how to control the fluid and air. I always test my fan pattern and finish delivery rate on scrap wood or cardboard so that I can make adjustments before I actually spray the piece.

Turbine-driven HVLP systems

Adjusting a turbine-powered spray gun is a simple process: no matter what type of gun you own, the idea is to start air flowing through the gun first, and then introduce finish slowly until it flows continuously and evenly. The gun should apply a full, wet coat with no heavy spots or misses. From this point, you can open or close either knob to obtain the best spray rate and fan pattern.

If you want to spray a lot of material in a hurry, open the fluid control more. If you are coating large surfaces, widen the fan pattern. If you're trying to achieve a fine finish or you're spraying small items, you'll have more control of how much finish is applied and where it lands by restricting the fan and fluid. But remember, how you set one knob affects the other. For example, if you increase the air flow without adjusting the fluid, the finish may be too fine. Conversely, opening the fluid control without widening

Check the finish with a viscosity cup. A stopwatch and the recommended viscosity cup show whether thinner must be added. Once thinned, the finish is passed through a filter.

the fan can cause runs and sags. At the ideal settings, the finish will coat evenly and flow together well.

Compressor-driven systems

With high-pressure spray guns and conversion-air HVLP guns (both powered by a compressor), you have the ability to control the air pressure entering the gun in addition to adjusting the fluid rate and fan shape. Getting all three adjustments coordinated can be a bit tricky and takes some trial and error, but being able to regulate the air pressure at the gun allows more spraying options.

Develop a spray strategy

Regardless of the size and shape of the object you're spraying, the main thing to keep in mind is that you want to spray an even coat over the entire piece. Always spray the finish in several thin coats rather than one heavy one. Lighter coats are less likely to run, dry faster and make sanding between coats easier.

If the pieces you are spraying are so small that the air from the gun blows them all over the place, try placing them on a piece of screen or wire mesh. I prefer spraying small parts with my turbine HVLP gun because the spray is softer. A good production tip for spraying many small pieces is to put them on a lazy Susan and spray several at once (see the top photo on p. 135). Rotate the turntable as you spray, so you don't build up too heavy a coat on the pieces.

Position large work on sawhorses or a stand so that the height is comfortable. You shouldn't have to bend, reach or otherwise contort your arm or body while you're spraying. You should be able to turn and move the work easily. I sometimes support the work on stickers or points (blunted drywall screws work well) to make sure that the bottom edge gets good coverage.

Spraying uniformly

To maintain even spray coverage, there are a few things to remember. Grip the gun firmly, but not so tightly that your hand gets

Spray the least visable areas first

Before spraying, make a dry run through the whole process. To help prevent you from overcoating or missing areas, visualize and then practice the sequence of spray strokes. Although the order in which you spray parts of a piece may vary slightly, there are a few rules of thumb worth following: Start with the least visible areas, such as drawer bottoms and cabinet backs, and work your way to those parts that will be seen. For example, spray the edges of tabletops, doors and shelves before the tops. This minimizes the overspray on the most visible surfaces.

Working from the inside out holds true for case pieces, too, as shown in the series of photos on the facing page. Always work from the wettest edge, so you can easily blend areas you've just sprayed. Where possible, move the gun away from your body, toward the exhaust fan (assuming you have one). This will help prevent overspray from settling on previously sprayed areas, and it will give you an unclouded view, too.

1. Spray overhead corners, and then fill in the inside top.

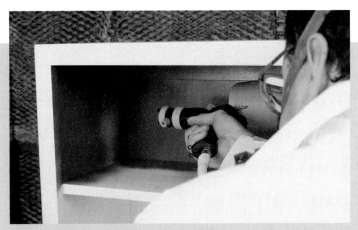

2. Coat interior back and sides. These areas won't be highly visible when the piece is finished.

3. Shelf tops and fronts—Remember to overlap strokes.

4. Finish the face frame. Begin with the inside edges, and then move to the front of the case.

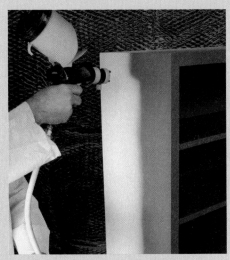

5. Do the exterior cabinet sides and front corners.

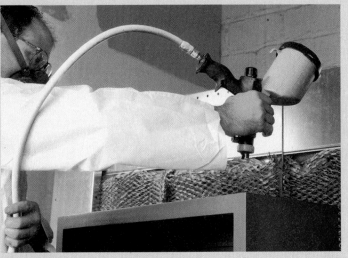

6. Spray the top. By leaving the top for last, the most visible part of the case isn't marred by overspray.

tired or uncomfortable. Point the nose of the gun so it's perpendicular to the work surface, and hold the gun at the same distance from the work on each pass. Move the gun parallel to surfaces, not in an arcing, sweeping motion. Begin your stroke 6 in. or so before the gun is over the wood, and continue the same distance beyond the other side. Trigger the gun a split second after you start your motion, and keep spraying until your arm stops. As you spray across the piece, move your arm steadily and smoothly without changing speed.

For most HVLP guns, hold the gun about 6 in. to 8 in. from the surface. This will let you spray a full, wet coat with minimal overspray and decent coverage. Move the gun at about the same speed you would a brush. Each pass should overlap the previous one by about half. When spraying small objects or tight places, reduce the flow and

move the gun closer. To avoid clouds of overspray and bounce back, work from inside corners out. Use more wrist action, and trigger more quickly. On large areas, increase the flow, pull the gun back an inch or two and make passes in opposite directions. I lightly spray across the grain to make a tack coat. Then I immediately spray with the grain.

In situations where your spray passes intersect, such as the stretcher-to-leg joint of a chair, release the trigger a bit sooner than you normally would. This will feather out the finish. If overlapping passes still give you a problem, mask off adjacent areas.

Drying and cleaning up are critical
It's easy to forget that once you spray a piece, the finish needs a warm, dry and dust-

Correcting spray-finish troubles

Fine Woodworking contributing editor Chris Minick found big improvements in his finishes when he switched to spray equipment. But the transition wasn't painless. Here's his list of common spray problems and, where they're not obvious, the solutions.

Orange peel

1) Atomization pressure too low: Increase pressure and adjust fluid.
2) Spray gun too far from work: Maintain 6- to 10-in. gun distance.
3) Coating viscosity too high: Thin to correct application viscosity.
4) Not enough coating thickness for proper flow.

Blush or cottoning

(Right half shows blush)
1) High humidity: Dehumidify shop, or add retarder to finish.
2) Improper thinner: Use only recommended thinner.
3) Moisture in spray equipment: Install water separator in air line.

White spots

1) Water contamination in spray equipment: Install water separator in air line.
2) Water on work surface: Dry work surface before spraying.

Sags and runs

1) Coat too heavy: Decrease fluid flow to spray gun.
2) Spray gun too close to surface: Maintain 6- to 10-in. gun distance.
3) Thinning solvent drying too slowly: Use faster evaporating thinner.
4) Drafty spray room.

free place to cure. If you don't have a separate drying area, production in your shop can grind to a halt. Even if you have a designated area, storing a number of wet cabinets, doors, drawers and trim pieces can be a problem. I use a system of racks to dry components and store them for short periods. Plywood trays, slipped into old baker's racks, come in handy when I have to dry lots of small pieces. When I'm drying round or odd-shaped items, like balusters, I hang them on an overhead wire from swivel hooks. Each piece can be rotated and sprayed and then hung in my drying area.

I clean my spray gun thoroughly while my work is drying. After cleaning the parts with the solvent recommended on the finish container, I dry them with compressed air. Then I coat the fluid passages with alcohol and let it evaporate before I store the gun in its case.

Turntable for even, quick coats—After arranging trophy bases on a lazy Susan, the author sprays with an HVLP gun.

Fat edge

1) Corner profile too sharp: Slightly radius 90° edges.
2) Drafts on one side of workpiece.
3) One side of workpiece warmer than other.

Cratering

(Solid chunk in center)
Solid contaminant (usually from non-loading sandpaper) lowers surface tension: Sand defect, and wipe entire surface with mineral spirits.

Fisheyes

1) Silicone or wax residue from paint stripper or old finish: Wipe surface with mineral spirits; mist coats (let each dry) to trap contaminants.
2) Oil in spray equipment (usually from compressor): Install oil separator in air line.

Microbubbles

(Haze, waterborne finish only)
1) Coating is drying too fast: Add retarder to finish.
2) Defoamer deactivates in waterborne finish: Don't use waterborne finish that's more than 1 year old.
3) Atomization pressure too high.

FIVE

Water-Based Finishes

It's proverbial knowledge that oil and water don't mix. Somehow, though, the chemists who have been developing water-based finishes forgot that bit of ancient wisdom. Instead, they forged ahead and found ways to suspend oil-based plastic resins in a water-based carrier, in spite of nature's best efforts to keep oil and water apart. Another proverbial bit of wisdom should also be questioned: that oil-based finishes are superior to their water-based counterparts. Top-notch water-based finishes have been a reality for some time. For example, you now can buy exterior-grade latex that's as good as or better than an oil-based paint. In fact some manufacturers recommend using certain latexes over an oil for topcoats.

Though the world of water-based furniture finishes has lagged behind house paint, it's now catching up. In a few years, water-based finishes will have moved unquestionably out of the minor leagues and be able to stand shoulder to shoulder with their oil-based counterparts. Outside of performance, they take oil-based finishes hands-down. They are much less toxic to both you and the surrounding environment (though if you spray them, a respirator is still an excellent idea). They don't need explosion-proof vent fans or fire-proof storage cabinets. And perhaps most important, the brushes and spray equipment are easy if not a pleasure to clean up—you simply use tap water, not paint thinner. So make the choice while you still can, for the days of oil-based finishes may be numbered. Clean-air legislation across the country keeps getting tougher, making it harder for manufacturing industries to use oil-based finishes. Depending on state and national legislators, in 20 years, oil-based finishes may go the way of lead paint and square-head jointers.

This chapter includes two articles: one on choosing among water-based finishes and another on how to use them. There's a lot available today, including water-based Danish oils, polyurethanes, lacquers, and even shellacs. You'll find tips on how they differ from oil-based finishes, how to work with them, and what to expect. Perhaps the most important difference is that water-based finishes tend to have a slight bluish cast. To compensate for this, you'll find a sidebar on tinting them with warmer colors.

NEW WATER-BASED FINISHES

By Andy Charron

Target Enterprises Oxford Hybrid Gloss Varnish
Good depth, warm tone. Closely approaches lacquer in appearance. Doesn't do well over oil-based stain.

General Finishes High Performance Polyurethane
Slightly cloudy, plastic appearance. Flows out well.

Compliant Spray Systems Enduro Wat R-Base Poly Overpri
Top-rated finish with a warm tone and good depth. Closely approaches lacquer in appearance.

A lot of woodworkers won't get near water-based finishes because they believe these products cause excessive grain raising, don't adhere well over oil-based stains and look like plastic. When I first began using water-based finishes about eight years ago, these products were indeed difficult to use and didn't look so hot. That's not the case anymore. Water-based finishes are getting better all the time. Also, they don't give off noxious fumes, they dry fast, and they aren't flammable.

A prior survey in *Fine Woodworking* evaluated 15 water-based finishes. Manufactur-ers have been busy, and there's a whole new crop of finishes on the market. I tried nine new finishes and compared them with a couple of time-tested finishes: nitrocellulose lacquer and shellac. I also compared the new finishes to Famowood Super Lac, a water-based finish that did extremely well in the previous evaluation, especially when measured on appearance.

The new finishes really stand out when it comes to stain resistance. Most were bulletproof. Grain raising wasn't objectionable with the majority of the finishes, and a few barely raised the grain at all. Some of the finishes were difficult to apply, although most went on without a hitch.

Rating the products on appearance is the most subjective test, but an important one, and several finishes

J.E. Moser Cool-Lac
The first water-based shellac. Has a warm color, but finish is very thin and doesn't build well, which makes it difficult to rub out without cutting through coats.

Parks Pro Finisher Polyurethane
Only a very slight blue tint (approaching neutral) and good depth. Fills pores well.

Sherwin Williams Commercial
spray finish, available only in 5-gal. containers. Slightly blue tint and cloudy. More difficult than average to apply.

Van Technologies VanAqua Urethane
Neutral tone but slightly cloudy. Builds and fills pores quickly.

FSM Corp. Clearly Superior 455
Slightly blue tint, cloudy. Fills pores well but difficult to rub to an even gloss.

Hydrocote Resisthane Pre-Catalyzed Lacquer
Slightly blue tint, cloudy. Easy to apply.

Product	Manufacturer
Clearly Superior 455	FSM Corp. (800) 686-2006
Cool-Lac	J.E. Moser (Woodworker's Supply) (800) 645-9292
Enduro Wat-R-Base Poly Overprint	Compliant Spray Systems (800) 696-0615
High Performance Polyurethane	General Finishes (800) 783-6050
Resisthane Pre-catalyzed Lacquer	Hydrocote (800) 229-0934
Kem Aqua	Sherwin Williams (800) 474-3794
Oxford Hybrid Gloss Varnish	Target Enterprises (800) 752-9922
Pro Finisher Polyurethane	Parks (800) 225-8543
VanAqua Urethane	Van Technologies (218) 525-9424
Benchmark finishes	
Solvent-based nitrocellulose lacquer	Various
Super blonde shellac	Various
Famowood Super Lac	Eclectic Products (800) 349-4667

Finishes were tested on mahogany plywood. The author applied an oil-based stain to half of each panel, then three coats of finish.

scored very high. Even the finishes that scored low on appearance are light-years ahead of what I was using five years ago. It's fair to say that water-based finishes are getting better, and I imagine the trend will continue.

Resins and additives have been improved

Traditional finishes such as lacquer and shellac have very few ingredients, primarily resins (solids that form the finish film) and solvents (also called carriers), which dissolve the resins. Water-based finishes are similar in that they, too, contain resins and solvents. But water-based finishes have many more additives than traditional lacquers, sometimes as many as 20, to deal with the basic incompatibility of water and resin. The other chemicals, especially ones called surfactants, allow water and resins to mix together, forming an emulsion. As the water evaporates, alcohols or cosolvents soften the resins, allowing them to coalesce and form the finish film.

Adhesion over oil-based stain	Stain resistance (22 max.)	Heat resistance	Raised grain	Sanding	Best applicator	Appearance
Pass	19	Fail	Moderate to heavy	Moderate	Brush or spray	Poor
Pass	22	Fail	Moderate	Easy	Brush	Good
Pass	21	Pass	Minor	Easy	Spray	Excellent
Pass	20	Pass	Minor	Easy	Brush or spray	Good
Pass	20	Pass	Heavy	Moderate	Brush or spray	Fair
Pass	22	Fail	Moderate	Moderate	Spray	Fair
Fail	22	Pass	Moderate	Moderate	Brush or spray	Excellent
Pass	22	Pass	Moderate	Moderate	Brush or spray	Good
Pass	22	Fail	Moderate	Moderate	Spray	Good
Pass	21	Pass	Minor	Easy	Spray	Excellent
Pass	16	Fail	Moderate	Easy	Brush or spray	Good
Pass	22	Pass	Moderate	Easy	Spray	Excellent

Although manufacturers are unwilling to give away trade secrets, they did tell me they've made headway with the types of resins and additives used in finishes. These improvements translate to finishes that bond better to solvent-based products and are tougher, yet easier to sand and rub out.

With the exception of one water-based shellac, the resins in the water-based finishes I tested are acrylic, urethane or a combination of both. These finishes go by many descriptive terms, including lacquer, polyurethane or varnish, but for purposes of discussion, most water-based finishes are technically lacquers, meaning they can be redissolved by their own solvents.

Finishes were all tested the same way

I used squares cut from the same sheet of mahogany plywood to test all the finishes (see the photo on the facing page). First, I applied a coat of Minwax red-mahogany oil-based stain to half of each panel. I allowed the stain to dry for two days, then applied three coats of finish to each panel, using

Finishes impress *FWW* editors

Three tables, three finishes. Walnut table has a nitrocellulose-lacquer finish. Cherry table is finished with Enduro Wat-R-Base Overprint. Maple table has an Oxford Hybrid Gloss Varnish finish.

To see how the two best-looking water-based finishes looked on furniture, we applied them to small tables. A third table was sprayed with nitrocellulose lacquer for comparison.

The Oxford finish on the maple has a lot of depth and makes the grain of the wood pop, although it has a heavy yellow bias. The cherry also looked good with a more natural tone. The lacquer actually seemed to have less clarity than the others.

—Matthew Teague, assistant editor, *FWW*

The Enduro and Oxford finishes both impart a nice, warm tone to the wood and are not muddy like other water-based finishes I've used. When I ran a fingernail across the three tables, the Oxford finish was fairly easy to scratch, unlike the other two. I'll try the Enduro on my next project.

—Anatole Burkin, associate editor, *FWW*

The three finishes are indistinguishable to me. They reflect light with the same clarity, and they all give a good sense of depth. Also, the water-based finishes don't have that bluish "skim-milk" cast to them.

—Jefferson Kolle, senior editor, *FWW*

either a brush, a spray gun or a combination of both, going by manufacturers' recommendations. I let each coat dry for a minimum of two hours before lightly sanding and applying another coat. Although some manufacturers offer sanding sealers, all of the products I tested can be used on bare wood, and that's what I did. Then I subjected the panels to common household chemicals to see how they would hold up.

After I had finished with the various tests, I sanded the topcoats using 240-grit, 400-grit and 600-grit sandpaper, then rubbed them with pumice and rottenstone. I also tested traditional shellac, nitrocellulose lacquer and a previously tested water-based product for comparison.

Testing adhesion over an oil-based stain
The adhesion test determines if a water-based finish will stick to an oil-based stain. I used a sharp knife or razor to slice an X into the finish where it was applied over the stain. Then I placed a piece of packing tape over the X and pressed down firmly. I let the tape sit for about five minutes and then yanked it off (see the photo below). The finish held fast on eight panels. Only one finish chipped off, failing the test.

Testing adhesion over an oil-based stain. A sharp knife was used to slice into the finish, then packing tape was placed over the cut. After five minutes, the tape was pulled off.

Improving the color
of water-based finishes

The color of some water-based finishes can be improved. To bring more warmth to a finish, dewaxed shellac can be applied as a sealer coat to bare wood. Or you can add universal tinting color (tube) or dissolved dye (bottle) directly to a finish to change its tone.

Although some water-based products have improved to the point where they appear comparable in color to traditional lacquer or shellac right out of the can, others still have a way to go. Some look a bit bland, and others suffer from a very slight bluish cast. If you happen to like everything about your finish except its color, consider toning the wood or the finish.

Some finish manufacturers tone their products for customers who prefer a warm look. The Enduro Wat-R-Base Poly Overprint has an additive that gives the finish a slight amber tone. Hydrocote offers an amber additive that can be added to its finishes. There are other products available to help you color topcoats. Whatever you do, don't thin the finish beyond what the manufacturer recommends.

Tinting the wood

Using shellac for a first, or sealer, coat will impart a warm glow to wood. It will also raise and stiffen the grain, making it easy to sand. Use fresh-mixed shellac that is dewaxed. You can also color the wood using dyes. A thin coat of a highly diluted water-soluble dye should give the wood just the right hint of color.

Coloring the topcoat

In some cases, you may want to color the topcoat itself. What you are actually doing is using the finish as a toner, which can be a bit tricky.

Pigments should be used in small amounts: You can use universal tinting colors (UTCs), which are available at paint stores, to alter the appearance of

a clear finish. A small drop or two of an earthy tone like burnt umber or raw sienna goes a long way toward giving an otherwise bland finish a sense of color and warmth. However, pigments are opaque and may give the finish a dark, cloudy or muddy appearance. If you use pigments to color a clear coat, use them sparingly and take great pains to apply the finish as evenly as possible.

Water-soluble dyes are preferred: A better alternative to pigments are water-soluble dyes. Dissolve a small amount of dye in water first and then add a few drops at a time to the finish until the color is right. Remember, water will thin the finish, so use it sparingly. Because dyes are transparent, they won't give the finish a muddy look. Dyes, however, will not penetrate the resins; they really only color the liquid part of the finish, which will evaporate, leaving the dye in place. This can cause some blotching.

Alcohol dyes are the best way to tint finishes: Dyes that have been dissolved in alcohol will actually penetrate the resins in a finish and change their colors. The resulting finish is even in tone and uniform in color. I like to use TransTint honey amber from Homestead Finishing Products (440-582-8929), which comes in concentrated liquid form. Add four to six drops per quart of finish as a starting point to impart a warm tone.

Add dye sparingly. A few drops of amber dye added to a neutral or slightly blue finish will give it a warm tone.

Food-stain test.
Finishes were subjected to a variety of common foods and cleaners. Most finishes held up very well. Cleaning products, however, were hardest on the finishes.

Common foods were used for the stain test

I gave each finish what I call the kitchen-table test. Tables are subjected to food spills, and a good finish should survive the onslaught as well as the chemicals used to clean up the mess.

I placed a small amount of the following common household products on each panel: milk, orange juice, hot coffee, mustard, ketchup, red wine, grape jelly, vinegar, olive oil, Windex and Fantastik (see the photo above). After one hour, I wiped each spot off, checking for damage. If there was no damage, the finish received 2 points. If the finish was slightly damaged (dull), it got 1 point. If the finish was severely stained, damaged or eaten away, it got no points. All of the finishes I tested scored at least 19 points, and several got a perfect 22. Windex and Fantastik caused the most damage.

A finish for a table needs to withstand heat

Tables are often exposed to hot pots, cups and spoons. To re-create this scenario, I put a spoon in boiling water for a few minutes, then placed it on a test panel (see the photo at top on the facing page). After the spoon had cooled, I removed it. If the spoon left no mark, the finish passed the test. If the spoon stuck to the finish or left a dull impression, it failed.

New products are easier to sand and rub out

Water-based finishes have a reputation of causing excessive grain raising and of being difficult to sand. A simple way to gauge the roughness is to brush a finger across the first coat after it dries.

The first coat of finish causes most, if not all, the grain raising. In my test, if a panel felt rough, like medium-grit sandpaper, I rated the raised grain as "heavy." If the panel felt more like fine sandpaper, I rated it "moderate." If it felt like very fine sandpaper, I rated it "minor."

To determine whether a finish was easy or difficult to sand, I sanded all three coats with varying grits of fresh paper. I not only looked at how easily the finish powdered up and how much it clogged the paper, but I also considered how hard I had to work to achieve a smooth, flat surface. Thankfully, all of the finishes I tested fell in the easy to moderate range. In fact, several of the finishes were as easy or easier to sand than lacquer and shellac. For a high-gloss, rubbed-out finish, I found that using 1000-grit and 1200-grit wet abrasive paper (available at auto-body supply shops) prior to rubbing out with pumice and rottenstone gave me the best results.

Appearance is the most subjective test

Because most of the finishes scored well in the kitchen-table tests, the deciding factor comes down to looks. Admittedly, this is subjective. Something that looks good to me may not look good to you. To give a fair evaluation, I showed the panels to a couple of other professional woodworkers and weighed their opinions as well as my own. I used solvent-based lacquer as the benchmark against which all the finishes were judged for appearance. Lacquer imbues wood with a warm tone, what I call a light amber color, and the finish also has clarity, which adds depth, especially after three coats or more. The best of the finishes approached this look. I downgraded finishes if they had a slightly cool or blue cast and were dull or cloudy.

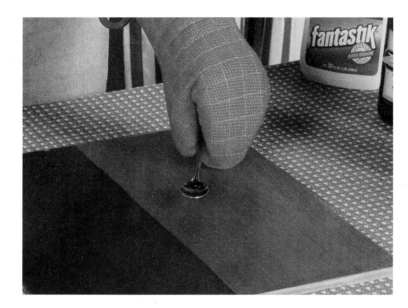

Overall, the new products tested well
If the color of the finish were not an issue, I would consider using any one of these products, with the exception of J.E. Moser's Cool-Lac. Although the Cool-Lac scored well on the tests (better than traditional shellac) and had a true shellac color, I found it difficult to apply. The product is very thin, nearly the consistency of water, and contains a low level of solids. As a result, the coats go on very thin and don't build well. Building a deep, protective finish requires half a dozen coats or more.

My two favorite finishes are the Compliant Spray Systems Enduro Wat-R-Base Poly Overprint and Target Enterprises Oxford Hybrid Gloss Varnish. Enduro and Oxford look virtually identical to solvent-based lacquer. They give wood a warm tone and highlight the grain because of their clarity. Both are easy to apply, but the Enduro requires only three coats to develop a good build. When I went to rub out the Oxford after three coats, I cut through to bare wood before achieving a nice shine. You need to apply at least five coats if you plan to rub this finish out to a high gloss.

If I had to choose between the two, I would opt for the Enduro, because it dries incredibly fast and because it passed the adhesion test. I think most people would be hard-pressed to tell the difference between Enduro and solvent-based lacquer. The Oxford finish—which, like the Enduro, has a nice, warm tone—didn't adhere well over an oil-based stain (it held fine over bare wood). You can solve this problem by using a water- or alcohol-based stain or by applying a sealer of shellac between the oil stain and the finish. Not to be forgotten, Eclectic Products Famowood Super Lac, a finish that tested well in the previous article, ranks right up there with the Enduro and Oxford. It's easy to handle and apply. It rubs out extremely well, has good depth and a color very similar to that of nitrocellulose lacquer.

USING WATERBORNE FINISHES

by Andy Charron

Before I owned any spray equipment, I used brushes or rags to apply solvent-based finishes. When I finally purchased a spray gun, I had a limited amount of money and very little shop space, so I could not set up a proper spray booth. I sought out finishes that were nonflammable and relatively safe to use. Waterborne lacquers were the obvious choice. All I needed was a fan for air circulation and a clean place to spray.

It took trial and error, but now I get consistently even coats of finish that are smooth and free of defects. I've also discovered that I don't have to use spray equipment to get good results. A number of waterborne finishes can be successfully applied with brushes or pads. Even though I now have the shop equipment to spray solvent-based lacquers and varnishes, I use waterborne finishes 90% of the time.

Many states now regulate the amount of solvent or volatile organic compounds (VOCs) that may be released into the air by professional shops. This has led to the development of more user-friendly and less-toxic waterborne finishes. However, waterborne products are still very different from their solvent-based counterparts. If they are not applied properly, they can be frustrating to work with and can yield disappointing results. Knowing what problems to expect and understanding how to overcome them will help make waterborne finishes easier to apply.

Success depends on several factors: surface preparation, compatibility of sealers, stains and topcoats, material preparation, application methods and even the weather. My methods are applicable to waterborne urethanes, lacquers, enamels, dyes, sealers and primers.

Prepare the surface
by raising the grain

If you have ever spilled water on a freshly sanded piece of wood, you may have noticed how the grain stands up, creating a rough surface. All waterborne finishes have this effect on wood. Earlier versions contained more water than the newer formulations, so grain-raising isn't as bad as it used to be. The resins used today are lighter, more viscous and require less water in their formulations. But no matter how much you sand bare wood, all waterborne finishes will raise the grain at least enough to require some additional sanding (see the top right photo on p. 148).

The simplest way to deal with raised grain is to surrender to it. First, finish-sand workpieces as you normally would with a sandpaper in the 180-grit to 220-grit range and then intentionally raise the grain. You can use water, sanding sealer or dewaxed shellac. If you use water, lightly dampen a sponge or a rag, and wipe the workpiece. Or you can dampen the wood with a plant mister. Let the workpiece dry to the touch, and then sand with 220-grit to 400-grit paper. A waterborne finish, when applied over this surface, will not raise the grain very much. A light sanding after the first coat is required, but you would be performing this step when using a solvent-based finish, too.

I usually raise the grain with a coat of sanding sealer instead of water. Most manufacturers offer sealers that are designed for their products. Sealers are usually formulated with stearates, which act as lubricants and make sanding easier. If you can't find a sealer, shellac works very well.

If the wood needs to be colored, I use one coat of water-soluble dye to raise the grain and then follow with a coat of sealer or shellac. When that dries, sand it. The sealer or shellac stiffens the fibers raised by the dye, making them much easier to sand. The sealer also gives you a buffer that keeps you from sanding through the dye to bare wood so quickly.

The amount of grain raised will vary with the type of wood. Open-grained woods, such as oak, will require more sanding than closed-grain woods, such as maple. I use wet-or-dry sandpaper in the 220-grit to 400-grit range, depending on how fine a surface I'm after. I don't use sandpapers that contain stearates. Small stearate particles that aren't cleaned off the workpiece surface will cause surface defects called fisheyes

Waterborne finishes will raise the grain.
Apply a sanding sealer over a stain or dye
before any topcoats. Sanding sealers contain
lubricants, which make them easy to sand.

Don't use tack rags to wipe off dust. They
can leave chemical residues that will show up
as blemishes under a waterborne finish. Use a
rag dampened with water.

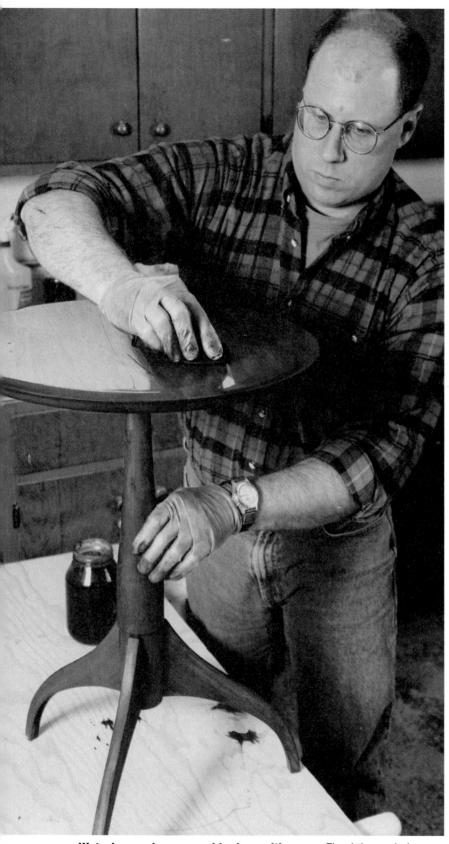

Waterborne dyes are rubbed on with a rag. Flood the workpiece
when applying stains and dyes. Work quickly, and wipe off any excess
to avoid lap marks.

when waterborne finishes are applied over them. After sanding, use a slightly damp, lint-free cloth to wipe off the dust (see the bottom right photo on the facing page). By the time you get out your brushes or set up your spray equipment, the workpiece will be dry enough for a finish. Do not use tack rags because the resins in them can react with the finish and leave blemishes.

Make sure all finishing products are compatible

Waterborne topcoats must be compatible with any other fillers, stains or dyes that are applied. Most waterborne materials have improved and many are now compatible with solvent-based products. That does not mean that all materials will be compatible in all cases.

If, for example, you plan to apply waterborne lacquer over pigmented oil stain, give the oil-based product enough time to cure fully. Before applying the waterborne product, rough up the surface with a very fine-grit sandpaper so the first coat has a better chance to bite into the stain. Sometimes, two products demonstrate their incompatibility immediately and the topcoat will bead up or not flow out. Problems such as blistering can manifest themselves several days later. If you're unsure about compatibility, experiment on a piece of scrap.

The best way to eliminate any doubt about the compatibility of two products is to apply a barrier coat of sealer between them. The best sealer I have found is dewaxed shellac. Although you can buy shellac that has the wax already removed, often referred to as blond shellac, it can be hard to find and usually comes in large quantities. I buy clear, pre-mixed shellac in a 3-lb. cut and keep it undisturbed for a day or two until the wax settles to the bottom of the can. Then I pour off the clear, top fluid. I thin it down to 2:1 with denatured alcohol. Then I apply a fairly heavy, even coat of this, let it dry for about a half hour and lightly sand with 220-grit (or finer). The shellac not only seals in the first coat but helps the two potentially incompatible materials bond. It's never failed for me.

Thoroughly mix and strain finishing materials

Most waterborne finishes are designed to be used straight from the can and do not require thinning. The only thing you need to do before applying them is to stir up the solids that settle to the bottom of the can. These solids have a tendency to separate or settle out over time and may require a lot of stirring to get back into solution. The older the material, the more likely it contains lumps. As a final precaution, I always strain it through a plastic, paper or nylon-mesh filter (see the photo on p. 151).

Occasionally, you may need to thin a finish such as a thick, pigmented primer because it doesn't flow or spray well. Unlike traditional nitrocellulose lacquers, which can be thinned almost indefinitely, waterborne finishes are extremely sensitive and don't respond well to thinning. Waterborne materials contain carefully measured amounts of various chemicals including solvents, water, defoaming agents and resins. Adding another material to the mix can upset this balance. When that happens, the finish may be prone to runs and drips because it takes too long to dry.

If the finish isn't flowing out properly after brushing, check with the manufacturer to see if a flow-additive is available. As a last resort, try adding small amounts (3% to 5% by volume) of clean water. Ideally, you should use distilled water, but I have used plain tap water without any noticeable ill effects. If the finish seems to go on too dry when spraying in hot, dry conditions, you might want to add a retarder (the surface will look and feel fuzzy).

Choose an application method

There are differences between waterborne topcoats made for spraying and those meant for brushing or padding. A spray finish is just that. If you try brushing it, the material may foam or dry too quickly. But I've found that any finish made for brushing can be sprayed with good results.

Brushing the Finish

Let the excess finish drip off the brush. Rubbing the brush against the edge of the container may cause the finish to foam.

Other causes of foaming—If you shake a can of waterborne finish instead of stirring it, you'll have a problem with bubbles.

Once you've started, work from a wet surface to a dry section. Brush quickly and with the grain; let the bristles skate off the workpiece surface to lessen brush marks.

Get the lumps out.
Waterborne finishes
have a high solids
content, so it's
important to strain
the material before
spraying.

Most waterborne stains and dyes don't require any special application equipment and can be wiped or sprayed just like solvent-based stains. However, because waterborne products dry so quickly (in particular, water-soluble dyes), you will have to move rapidly when wiping them on. Be sure to flood the surface with a full, wet coat to avoid lap marks.

I usually get a good finish with two applications of topcoat. For added durability, such as you might need on a tabletop, I'd recommend three or more coats. Although waterborne finishes don't release the kind of noxious fumes some solvent-based finishes do, they still give off some vapors. So I take precautions. If I'm brushing finishes, I make do with some cross ventilation. When I'm spraying, I wear a respirator with organic vapor filters and ventilate the work area.

Select a synthetic bristle brush for finishing

Natural bristles will absorb the water in waterborne products and begin to splay and lose their shape. Synthetic bristles won't. When applying a finish, keep the brush wet, and don't scrape the bristles against the edge of the can (see the top left photo on the facing page). Let the excess material drip back into the container. This takes a little longer, but it will help prevent foaming. Then apply the material on the workpiece in a thin coat. Put it on too thick and you will get runs and sags. Always work quickly and from a wet edge to avoid lap marks (see the bottom photo on the facing page).

The more you brush the finish, the greater the likelihood it will begin to foam and bubble. If you experience foaming, add a flow additive for the finish, if one is available. If not, as a last resort, try adding a few drops of lacquer thinner, mineral spirits or milk to the finish. These additives can reduce the surface tension of the finish and improve flow. Disposable foam or sponge brushes and paint pads also work with waterborne materials. Apply the finish over the surface using quick, light passes.

Spraying gives the best results

A spray gun allows you to apply a full, even coat over an entire piece in a manner of minutes. The finish dries so quickly that, in most cases, you will be able to apply several coats in one day.

Because waterborne finishes contain a higher percentage of solids than most other finishes, they have a tendency to run or sag if applied too heavily. When spraying, lay on just enough material to leave a shiny, wet sheen on the surface of the wood, but not so wet that it begins to run.

If you catch a run or drip while it is still wet, wipe it off with a clean, lint-free cloth, and recoat the area immediately. Otherwise, use a razor blade to cut off any dried or skinned-over trouble spots, sand and recoat.

Spray equipment that's made of plastic or stainless steel is best for use with waterborne products because those materials won't rust. But if your gun is made of metals that can corrode, you can ward off rust by drying it thoroughly after use by blowing compressed air through it. You can also remove any residual water by running a few ounces of denatured alcohol through the gun.

Weather conditions affect finishes

The cooperation of Mother Nature can certainly make a difference when applying finishes. When waterborne materials are applied on dry, warm days, they flow out smoothly, level quickly and dry to the touch in less than an hour, sometimes in a matter of minutes when spraying. Under ideal conditions (around 70°F with 35% to 50% relative humidity), you can apply several coats in one day. However, if your finishing room is cold or the humidity is high, waterborne products can become downright ornery.

When waterborne products are cold, they don't atomize properly, don't flow out well and take longer than normal to dry. Ideally, you should heat your finishing room. But there's another way. I've found that if I heat the finish to about 75° right before using it, I can apply topcoats in a room as cold as 45°F. All I do is place the can of finish in a sink or bucket full of hot water for a few minutes. (Never use a stove or open flame to heat any kind of finish material.) Warm finish is easy to spray, flows out well and dries quickly.

Lowering the humidity can be more difficult. In a small room, a dehumidifier can reduce the moisture content. But I have a large shop near the ocean and no equipment to reduce humidity. I have found that a fan blowing warm air over the piece being finished can offset the negative effects of high humidity.

Waterborne finishes, like other topcoats, can be rubbed out to increase or decrease their sheen. Just remember to avoid steel wool, which can cause black spots if pieces of it lodge in the finish and rust.

Spraying the Finish

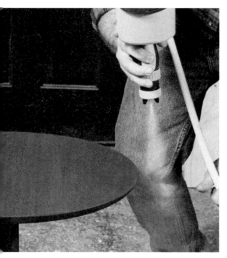

Begin spraying before you reach the workpiece. Hold the gun 4 in. to 6 in. away from the workpiece, and spray at a speed that makes the surface wet and shiny but not runny.

Don't stop before the edge. Keep spraying until the pattern falls off the edge of the work-piece. On the next pass, overlap the previous section.

Many waterborne finishes look milky white when first applied. The section closest to the author already shows signs of clearing up as he works toward the center of the table (above).

SIX

Special Techniques

This chapter on special techniques addresses particular finishing problems or desired effects. These are finishes that you won't use every day, but will be very handy when the occasion presents itself. And if any finishing doesn't contain a modicum of the unusual or "special" circumstances, then wood must not be involved. The special aspects of these finishes are really in the author's particular use. Read between the lines, and you'll pick up a whole lot of general finishing information applicable to everything you do. Then again, you might just pick up a good solution to a single, but important, problem.

For example, if you haven't already, one day you will attempt to finish cherry. If you reach for a clear Danish oil, you won't notice anything particular about it. Cherry will take on a warm lovely glow as do most other woods, and that will be that. But if by chance you pick up a pigmented oil or a stain, then watch out. Some cherry has a bad allergy to pigments and dyes. The symptoms are serious blotching (an uneven distribution of the pigments on the wood) and a refusal to be cured once blotchy. In the 18th century, this was one of the reasons cherry was considered a secondary wood, acceptable for parts hidden inside a carcase, but not for anything that showed (how times change). In a short article, Jeff Jewitt offers an excellent solution to this age-old problem, and one that will help you build fearlessly with one of America's most popular hardwoods.

Other articles address painting furniture, restoration, and refinishing problems. While it can be a difficult matter getting a good finish down on bare wood, it's a whole other bag of tricks fixing a pre-existing finish. It's an art that requires you to match colors and effects and leave no trace of the additions. Pinchas Wasserman and Robert Judd will show you how. And what do you do when people are going to put your next woodworking project in their mouths? Wooden spoons, bowls, and cutting boards shouldn't be finished with something that could be toxic if ingested. Jon Binzen explores the possibilities.

Finally, no good book on finishing would be complete without giving some space to opposing points of view. Pat Warner makes a case for not finishing wood at all, but letting it age and wear on its own. You might be thinking of arguments against the idea already, or agreeing with him outright as you realize all the time you'll save, but don't make any decisions until you listen to his arguments. You might be surprised.

FINISH CHERRY WITHOUT BLOTCHES

by Jeff Jewitt

Y ears ago, one of my first projects was a simple Shaker-style table for my wife. After carefully selecting the best grain and figure for all the parts, and sweating through the construction details, I was ready to apply a finish. I wanted something really special, so I chose a dark-red dye stain and applied it first to the tabletop. I had a sense something wasn't quite right as soon as I wiped the stain with a rag. Dark, ugly splotches began to appear. I was nervous, but figured the problem would disappear when the stain

dried, so I stained the rest of the table. The result was a disaster. I managed to salvage the top by removing a good ⅛ in. of wood with a belt sander and applying a clear finish. But those beautiful book-matched legs, covered with the most delicate ray fleck patterns, now stand in a corner of our living room under a coat of green milk paint.

Cherry is a joy to work. It's easy to cut, shape and sand. If left unstained and coated with a clear finish, it eventually matures to a deep, reddish-brown color coveted by

A clear coat
of thinner tells all

Knowing where splotching is likely to develop is
half the battle. By saturating the surface of a
piece of cherry with thinner (see the photo at
left), you can get a good reading on how splotch-
prone that wood will be.

Quick and Easy Method

1

Finishing doesn't get any easier than this. A sealer coat of oil (1) and subsequent top-coats (2) of a dark-toned shellac (button or garnet) will give a rich finish on cherry that will only improve over time. The author pads on the shellac with a lint-free rag.

2

Instant Aging Method

1

2

Start with a light amber dye stain. A first coat of heavily diluted water-soluble dye stain evens out the colors among different pieces of wood (1), and is the first step toward building up layers of color in the cherry.

This thick, gloppy stuff adds more color. Using a gel stain as a glaze (2), applied over a sealer coat of shellac, is an easy method for adding more color to the cherry.

antique dealers as well as woodworkers. In an attempt to duplicate that old-timey color and overcome the pink-salmon hues of freshly machined cherry, many woodworkers use full-strength stain as a first coat and end up with the same blotchy mistake I did. But there are ways to avoid splotching. The first step is to understand the causes.

Knowing why cherry splotches will help you avert the problem

Splotches develop in cherry (and other woods like birch, red alder and soft maple) because of the uneven penetration of stain. It penetrates unevenly for a number of reasons, and any one or a combination of them can condemn your finishing efforts. So before devising a strategy to prevent blotching, it helps to identify which obstacles may exist within the wood you have on hand. I know of three reasons why stain will penetrate unevenly in cherry.

Resin deposits, the most common culprit

Cherry is one of many woods that often have unseen, concentrated deposits of resin within the wood as a result of the kiln-drying process. The resin deposits attract stain solvents, causing stains to penetrate more in some areas and less in others. It's hard to know when this is going to happen. One easy test will warn you of trouble ahead (see the box and photo on p. 157). Saturate the wood with any common solvent, such as denatured alcohol, paint or lacquer thinner. Splotch-prone areas will show up right away because they will absorb the solvent faster, just like they would with stain.

Alternating grain, avoid it when possible

When the grain direction changes within the same board, the stain will penetrate unevenly. This effect can be dazzling, as in curly figure, but more often than not, as in areas around knots, a less-than-attractive appearance is the result. It's usually easy to avoid this condition when it exists, simply by reading the grain direction on the edge of a board and cutting around problem areas when you select wood for cabinet or furniture parts.

Nearing the finish line. When wiping off excess gel stain with a rag (3), you can control how much is left on the surface for just the right effect. Successive coats of orange or garnet-colored shellac build up the color in layers (4). The more coats you apply, the darker the finish will become.

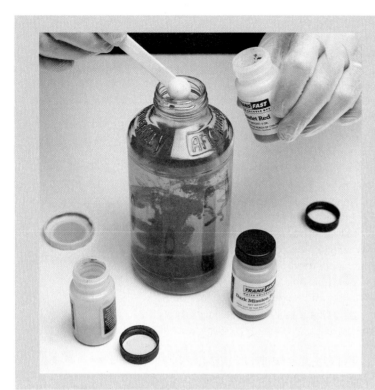

Mix your own dye stains

If you don't want to buy a premixed water-soluble dye stain, it's easy to mix your own from pure colors—by weight—using five parts lemon yellow, one part red and one part dark brown. If you don't have access to a scale, you can make a diluted 1-qt. solution by adding 1 teaspoon of yellow, a pinch of red and a pinch of dark brown using the tip of a spoon (see the photo at left). Mix the stain in a glass or plastic container, and wear gloves and a dust respirator when mixing dyes: This stuff stains skin as well as it stains wood.

Improper sanding, the easiest cause to detect and correct

The most obvious truths are sometimes hard to see: Improperly sanding the surface of any wood species can cause problems. And careful sanding is especially essential with cherry. Dull sandpaper can burnish the surface rather than cut it, making it less likely to accept a finish. Leaping from rough to really fine abrasives is also a no-no, leaving scratched surface areas that are more porous than others. A scraped or planed cherry surface will usually accept a clear finish evenly, but a stain applied over these surfaces will often spell trouble, too.

I usually sand cherry with a random-orbit sander, starting with 100-grit and proceeding up through 180-grit, changing grits at 120 and 150. I switch to fresh paper often and inspect the surface in backlighting to make sure I don't miss any spots. I then hand-sand, using 180-grit, with the grain of the wood. Such careful sanding won't eliminate splotching, but it will help to minimize it, especially in concert with the finishing techniques outlined below.

Two strategies that work

Some people attempt to tame splotching by controlling how much stain is absorbed. I've heard several woodworkers who swear that the new gel stains help control stain penetration, but I've found that they don't work very well on raw cherry. Washcoating is another technique popular with professional finishers. A washcoat seals off the surface of the wood with a very thin resin—diluted shellac, thinned oil or glue size—which decreases the penetration of a stain. But this technique is hard to control evenly, and less penetration means a lighter shade of stain.

I've used two methods to prevent blotches that are well within the range of just about any woodworker's finishing talents. Both are applied by hand, and both yield finishes that have depth and luster with little or no splotching. The first technique (see the top photos on p. 158), which is the easiest, will change cherry from its initial pinkish tone to a golden color that will continue to darken with age. I recommend this technique only for projects that have been carefully matched for grain and figure because exaggerated color differences in the lumber won't be concealed.

The second technique, which is a bit more complicated, will satisfy those who want a dark, rich color without waiting for nature to do the job. It can also be used for projects that are made from wood of varying color and figure.

The quick and easy method uses oil and dark shellac to supply color without stain

After sanding your project through 180- or 220-grit, apply a light coat of boiled linseed oil or Watco Danish oil. It's not necessary to flood the surface, just apply enough oil with a rag to make the wood look wet. This step enhances the grain and adds depth. After a day or so of drying time, lightly scuff-sand the surface using 320-grit sandpaper. Wipe off the dust with a rag, and then apply a dark garnet-colored shellac to the surface.

I use a 2-lb. cut (meaning a ratio of 2 lbs. of dry shellac flakes dissolved in a gallon of alcohol) and wipe it on with a rag in a process called padding, but you could also use a brush. This shellac has a dark golden-brown tone that adds a bit of color to the cherry. If you want a darker color, you can apply another coat of the dark shellac. If you're satisfied with the color after one application, you can apply a lighter colored shellac for additional coats. (For more protection, you can apply a varnish as the top-coat.) Build the finish to your liking, and after the proper drying time, rub out the shellac (or varnish) with 0000 steel wool and a dark paste wax.

The color of this finish will start out as a light golden brown but will quickly pick up reddish tones after a month or so when exposed to light and air. By the end of the first year, your project will have a deep-red tone that just keeps getting better with age.

Instant aging is trickier

This finish will result in a rich, dark color (see the bottom right photo on p. 159) and will help to even out tonal disparities in the wood. Prepare for the finish by sanding the wood through 180-grit. Using distilled water (free of minerals that might stain the wood), wet the surface of the wood to raise the grain. After the wood is dry, smooth down the raised fibers with 220-grit.

To color the wood and minimize splotching, you can build up color in layers. Apply a light amber-brown dye stain (see the bottom left photo on p. 158) as the first coat of color (called a base or a ground stain). This color is sometimes sold as honey amber, but you can use just about any light shade of golden-brown dye. The dye is diluted with four or five times the recommended amount of water and should have the appearance of strong tea. If you cannot find a suitable color of dye stain, you can mix your own (see the box on the facing page). This base stain, applied as a diluted solution, will add depth and will even out the tones of different boards.

After the base stain is dry, very lightly scuff-sand the surface with a synthetic abrasive pad. Seal in the dye with a 1-lb. cut of shellac, and let that dry. To add more color and depth, you can apply a dark pigment glaze. I use a dark-brown gel stain (see the bottom right photos on p. 158), such as Bartley's dark-brown mahogany, brushed on and wiped clean. After the glaze is fully dry, apply orange shellac, brushed or padded on, to add more color to the final finish. Keep in mind, no matter what finish you put on cherry, it will continue to darken all by itself, getting better looking with each passing day. The more it's exposed to light, the faster that will happen.

MAKING WOOD
LOOK OLD

by Jeff Jewitt

Match the tool marks

Period tools match the surface texture. Handplaning with a re-ground blade in an old jack plane produces the same pattern left by a scrub plane on the original hutch.

From a magnificent specimen of Cuban mahogany to a humble piece of white pine, wood looks better as it ages. All woods mature with use and time, developing the patina so valued in antique furniture. In my conservation and restoration business, I need to match the look of old wood to new when I'm fabricating missing parts for antique furniture.

I try to simulate the order in which the wear and tear would have happened naturally. I start by matching the surface texture of the new wood with the old. I follow that with a dye stain, distress marks and glazing coats to add more color. Then I apply a finish to match the original.

Match the original surface texture first

Furnituremakers of two centuries ago prepared wood differently from the way we do it now. Lumber was dressed, shaped and made ready for finishing solely by hand. Their tools left distinctive marks on the wood very different from those left by modern milling and sanding methods. Edges and moldings were shaped with molding planes or carved with gouges and chisels. After planing, surfaces that would show were smoothed and evened out with scrapers or glass paper (made by pulverizing glass and sifting it over glue-sized parchment).

Even on some very fine, more formal antique furniture, you'll often find marks from tools such as rasps and files that were used to clean up the ridges left by sawblades and chisels. Molding planes produced crisp, deep profiles unattainable with many modern shaping bits. Although results may seem somewhat crude by today's standards, the goal then was the same as it is now—to produce as flawless a surface as possible.

Flat surfaces on many country-style antiques have a slightly scalloped appearance produced by fore planes, or scrub planes, and scrapers. The scallops are readily apparent under a finish and when viewed in backlighting. To re-create this effect, I ground a very slight convex profile on the blade of an old jack plane (see the photos on the facing page), making sure to relieve the corners of the blade. A very small relief is all that's necessary. Flexing a scraper with your fingers will create a similar pattern. When you use any of these tools, small tearouts or other imperfections in the wood—a sign of handwork—should be left alone.

Patina is more than an old finish

Patina is the surface appearance of something that has grown beautiful with age or use. The much desired patina on antique furniture involves alteration of both the surface color and the texture of the piece as it ages.

Wood contains natural dyes and pigments responsible for the characteristic color of each species. A change in color, a result of exposure to light and air (photo-oxidation), may be the most obvious effect of age. As a rule, light-colored woods darken, and dark-colored woods lighten.

Another kind of patina develops as stains and finishes age and as wax builds up on the surface of the wood. Photo-oxidation causes dyes and pigments to fade and finishes to yellow and darken. Over the years, polishes and waxes build up in corners, cracks and crevices and act as a magnet for dust, which accumulates on surfaces that are not regularly cleaned.

Most old furniture ends up soiled, dented, scratched, eaten by insects or worn-out from normal use and handling. Oils from skin produce a grimy buildup around hardware and other areas where furniture is handled. The bottoms of legs get banged up the most. Sharp edges and corners that are regularly handled become rounded. Everyday contact with clothes and shoes will eventually wear finishes and stains down to bare wood.

Use dyes, bleach and light to change the color of wood

You can duplicate the effect of light on wood with either dyes or chemicals. Both produce a color change within the structure of the wood. Although their effects are similar, one very subtle difference is that dyes tend to accentuate figure and grain and chemicals do not. Dyes are, by far, easier and safer to use. They can be soluble in alcohol or water. Some alcohol dyes are extremely light-sensitive, and they will not hold their color over time. Water-based dyes tend to be less vivid than alcohol dyes and produce a more believable color. Although water-based

Test the stain on scrap first. The author used a cutoff from the new pine shelf to fine-tune his custom-mixed dye. By adding small amounts of red dye to his initial mix, he was able to get a better match.

Four hours in the sun show a dramatic color change in this piece of cherry. The center of the board was covered with duct tape to keep the light out, and the right half was coated with thinner to approximate a clear finish.

dyes raise the grain in wood, producing a rough texture, the problem can be minimized by applying a wash coat of plain water and sanding off the fuzz after the surface dries before the dye goes on.

In almost all light-colored species, a yellowish-brown dye stain will simulate the color of aged wood. This stain color is sold pre-mixed by many companies, often called honey or amber, but you can make your own from powdered dyes in primary colors. The formula I use most is 10 parts lemon yellow,

one part red and one part black by weight, not volume. Use this color on birch, maple and oak. It also works well for warming up the cold tones of kiln-dried walnut. Used on Honduras mahogany, it will kill the pink tone in preparation for subsequent coloring layers. With one or two more parts of red added, a nice aged pine color is the result (see the photo above).

Some dark woods—rosewood, teak, walnut and old Cuban mahogany—lighten considerably after being exposed to light for a long time. To match these woods, you may need to bleach the new wood first and then treat it with a dye stain.

Use a two-part bleach of sodium hydroxide and hydrogen peroxide. Avoid using any other chemicals on woods that have been bleached: A chemical reaction may create harmful fumes.

Some woods, such as poplar and cherry, darken considerably after only limited exposure to sunlight (see the left photo on the facing page). Cherry will darken in ambient room light after a few years to a dark, reddish orange. To hasten this process, finish it with a light coat of boiled linseed oil followed by the finish of your choice. After several months, you will have a color that would normally take longer to achieve.

Distressing: when and how to alter the surface

You can imitate dents by using the tang of a file after the first coat of stain but before the glaze goes on. Scratches can be made with a piece of glass or a wire brush (see the top photo at right). Very small drill bits and the point of an awl will mimic worm damage (see the bottom photos at right). Drill the wormholes after the finish has been applied but before the last coat of tinted wax.

To wear away edges, wrap some thick twine or thin rope around your fingers, and pull it back and forth, shoe-shine style, across the edges of tops and stretchers. To round off corners, use a brick, and then burnish the wood smooth with a piece of hard maple. Anything goes, except overdoing it. Too much wear will look contrived.

Finish the job with a glaze

The best way to duplicate the depth of color in old wood is with a glaze. Glazes are thin, transparent layers of color applied over another color. Before applying a glaze, it is best to seal in the first layer of color with one or two coats of finish. I prefer shellac. To match most old furniture, the best glazing colors are brown, umber and sienna—sometimes called earth colors. You can use a pre-mixed glaze, or you can make your own if you want better control over the color.

I prefer to use a clear glaze medium (like Behlen's heavy-bodied glazing stain, which is thick and has a long open time). I tint the

Add some wear and tear

Wire brushes abrade the surface and mimic the wear and tear of two centuries of use. Glass can also be used to make similar scars.

Counterfeit wormholes—After applying a shellac finish but before the final coat of dark, tinted wax, the author uses a small drill bit and the point of an awl to match damage done by worms.

glaze with dry pigments. Unlike dyes that dissolve into water or alcohol, pigments are suspended in the glaze medium. I also tint with Japan colors, a kind of concentrated paint that will mix easily with oil-based products.

Normally, I do all the distressing before I apply the glaze because the glaze will collect in dents and scratches and provide a very convincing effect. I brush the glaze over the

Apply a glaze to add depth

A glaze adds depth and color. After applying a glaze with a brush (left), the author controls the amount of color left on the surface as he wipes it off with a clean rag. When he highlights some areas more than others (above), he dabs on the glaze selectively and blends it in with a dry brush. This technique will also add extra color to distress marks.

Final adjustments before the finish goes on. The author wedge-fit the new shelf into this antique dry sink and added another coat of glaze to adjust the color. When the color was right, he removed the shelf and applied a sealer coat of shellac.

entire piece (see the top left photo above) to add an overall color effect or selectively dab it in crevices and corners where wax is likely to build up (see the center photo above). The oils in glaze mediums never dry fully, so it's normal to feel some tackiness, even after several days. Glazes and dry pigments are available in many woodworking supply catalogs. Japan colors can be found at professional finishing suppliers or at some paint stores.

Glazes should be sealed with at least one coat of clear finish. It's best to spray on shellac or lacquer. If you have to use a brush, flow on a thin, 1-lb. cut of shellac without applying too much pressure. You can follow that with a varnish. The seal-coat of shellac is important because the varnish may not bond well to the glaze.

As a final step, I rub it out with steel wool and a dark wax, such as Liberon or Behlen's brown wax. My favorite is Antiquax brown wax. It's tinted with oil-soluble dyes and pigments and matches the look of built-up old wax beautifully. I apply the wax by unraveling a piece of 0000 steel wool and refolding it into quarters. I squirt some mineral spirits onto the pad, dip it into the can of wax and smear it all over the wood surface, working the wax into corners and distress marks. After the wax hazes over, I buff the surface with a clean cloth.

CREATING AN ANTIQUE PAINTED FINISH

by Kirt Kirkpatrick

No, it wasn't made by the conquistadors. Though it looks like it's been in a Spanish Colonial mission for several hundred years, this hall table is really less than a year old.

I started experimenting with painted finishes that look old because I live in a very old region of the country. The Native American and Spanish Colonial cultures are still very much a part of the look here in New Mexico.

In collaboration with my friend Dwayne Stewart, who's a painter and professional finisher in Kansas City, Mo., I've developed a method that makes even new furniture look like it's been around for a long time.

Selecting and preparing the wood

I use old wood whenever I can, but new wood can be stained dark to make it look older.

Tool marks make a big difference, too. I eliminate machine marks with hand tools, and I gouge the wood intentionally. A 17th-century Spanish craftsman here in the desert Southwest might have had an adze, a drawknife, maybe a handplane (but likely not) and not much more. And he certainly didn't have any fancy sharpening stones. So the surfaces you see on most old furniture around here is kind of rough. I achieve a similar effect by planing against the grain in places (especially near knots), causing tearout, skewing the blade on my plane so it gouges the surface, keeping the blade intentionally dull and burnishing sharp edges. This may run counter to everything you've learned, but the results are convincing (see the photo on p. 167).

Once I'm happy with the surface, finishing begins. Because I use latex paint and a quick-drying clear coat, I can complete the process in less than two days (see "An antique finish in 12 steps" on the following pages for a thorough description of the process). Not bad for a finish that looks like it's seen some history.

An antique finish in 12 steps

1. Burnish the edges. Furniture doesn't age, or wear, evenly. Sharp corners, edges and other crisp details soften first. The author uses the shank of a large nail to round over the sharp edges on a tabletop.

2. For a light wood like pine, use a dark stain. Because wood changes color as it ages, the author uses a pigmented oil stain (Minwax Early American) to darken this tabletop made of ponderosa pine. But any kind of stain will do. Then he lets the stain dry according to the manufacturer's instructions.

3. Seal in the color with a clear coat. The author brushes on two coats of lacquer, but other clear finishes will work as well. Just be sure to use something with a low sheen.

4. Scuff-sand the clear coat. A quick once-over with 220-grit dulls the sheen and gives the clear coat enough tooth to hold a coat of paint.

5. Wax prevents paint from adhering, which lets the stained wood show through. Rub a bar of paraffin lightly over the edge and a bit on the top. Let the bar skip along, so the pattern will be uneven. Wax the edge more heavily, but still intermittently.

6. Apply a first coat of flat latex paint. Coverage doesn't have to be perfectly even, and it's probably better that way. Choose a color that contrasts well with the topcoat. Give it an hour or two (or whatever it says on the can) to dry.

7. Brush on a coat of hide glue. The author uses pre-mixed liquid hide glue, but hot hide glue also works. If the premixed glue appears too thick to brush out, thin it slightly with some warm water. Mix well before applying it. A thicker coat will give you fewer, bigger cracks in the next layer of paint; a thinner coat will give you smaller cracks but more of them. Don't worry about laying down an even coat (variations in the size of the cracks look more realistic), but apply the glue in only one direction. If you're haphazard with your strokes, the crackle pattern won't look right. This is the only step you really have to be finicky about. Give the glue half an hour or so to dry.

8. Apply a second coat of flat latex. Make sure that the paint is flat; semi-gloss or gloss paint won't crackle. Keep a wet edge, move quickly and don't go over your previous strokes, or you'll fill in the cracks. This second coat starts to crackle almost immediately. Let it dry thoroughly, preferably overnight.

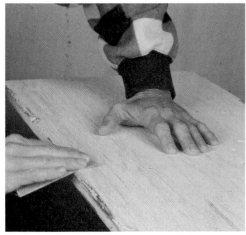

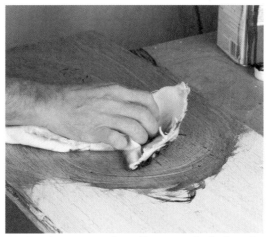

9. & 10. Scrape and then sand the top and edges. When the second coat of paint is dry, use a paint scraper to remove paint sitting on top of the wax. The scraper also will dislodge loose chunks of paint to reveal the first layer below. Mist the surface with water, and then rub with your fingers to create an even more authentic look. Sand lightly to soften sharp edges.

11. Apply a coat of medium- or dark-tinted liquid wax. The author uses Watco dark-satin finishing wax. This wax seeps into all the cracks and recesses and gives the whole piece a darker, almost dirty look—instant patina. Temperature affects drying time. The author usually waits about 10 to 15 minutes.

12. Remove most of the tinted wax with a clean rag. If the whole piece or just some areas are too dark, you can remove some of the color. Apply a clear coat of paste wax and rub vigorously. The solvent in the wax lifts the excess color from the surface. The paste wax protects the surface, too.

BETTER PAINTED FURNITURE

by Chris A. Minick

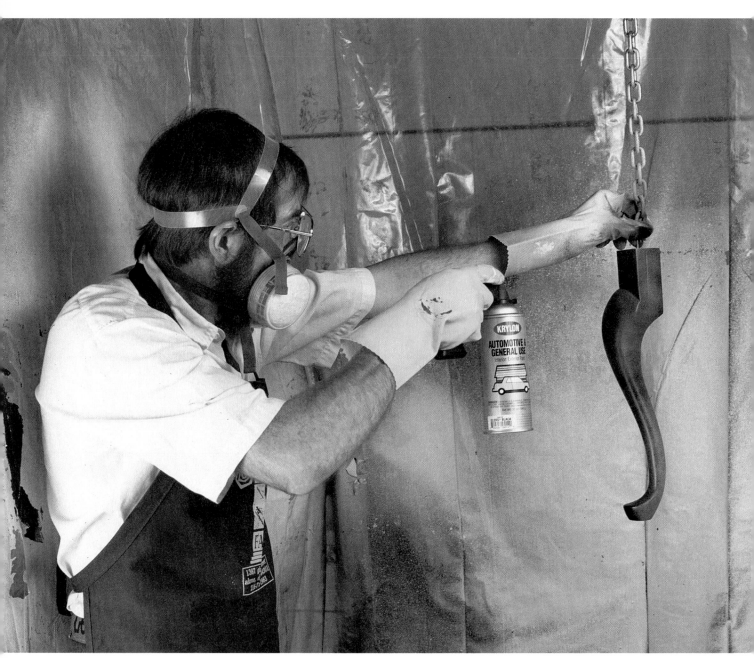

Automotive aerosol paint makes for a flawless finish. Wearing an organic-vapor respirator and rubber gloves, Chris Minick sprays a cabriole leg in a number of light coats. By hanging the leg from a chain, he can rotate the leg as he fills in missed areas. The spray booth is a U-shaped drape of 4-mil polyethylene hung from his garage rafters. Minick leaves the doors open for ventilation.

Paint showcases furniture's wood and form—Painted maple legs emphasize the lines of the author's highly polished coffee table. The black tabletop edges, bottom apron bead and corner legs make an attractive frame around the ribbon-striped-mahogany veneer top and quilted-mahogany veneer apron. The underside of the table is also painted.

"If it works, don't mess with it," sums up the attitude that many woodworkers have toward finishing. Learning about a new finishing technique can be complicated and confusing. So it seems easier to stick with an old standby like tung oil, or stain followed by varnish, even though it may be merely adequate. If that's your habit, you may have overlooked an important class of finishes—paint.

Paint is a versatile medium because it can be used as a design accent to emphasize the lines of a piece, or it can be used to draw attention to handsome woods in furniture (see the photo above). A painted finish also lets you use up those too-good-to-burn pieces of scrapwood. But don't be mistaken. Paint cannot cover up poor workmanship or shoddy surfaces. A painted finish requires better preparation than a clear finish. Fortunately, there are some products that make the whole process relatively painless.

If I have to paint a fairly large project or one that needs a special color, I use a good latex paint and an airless sprayer. But for most items, especially the ones that require a professional-looking paint job (such as the coffee-table leg in the photo on the facing page), I use ordinary aerosol spray cans for priming, painting and clear coating. Auto-parts stores have a marvelous variety of colors and types to choose from. And automotive fillers and putties are superb too.

Learning from automotive finishers

Folks working in the automotive industry are constantly refining paint finishes, due to the meticulous demands of car finishes (see the box on p. 174). That's the main reason I buy many of my furniture-finishing products, including fillers, primers and paints, from my auto-parts store. And given the fact that paint is more easily scratched and more difficult to repair than most clear finishes, I borrow another technique from automobile finishers: I clear coat my painted finishes. Before I buy anything for a project, though, I think through my whole painting strategy.

Planning your paint job

Painting, like any finishing technique, can be frustrating when some unexpected problem arises halfway through the process. The best way to eliminate surprises is to test all your materials and practice new techniques on scrapwood. After all, you wouldn't cut dovetails the first time using prized wood for your project. So you should treat paint-finishing the same way. Paint decisions for a piece of furniture must be made before the first board is cut.

Because my furniture pieces often combine painted elements as well as stained and clear-coated portions, it's easier to finish each component separately, then assemble them. Though this requires careful planning of the construction and care in final

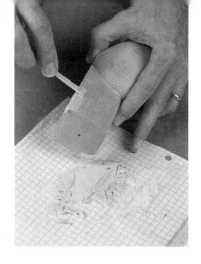

Two-part auto-body filler won't shrink, so it's great for leveling defects in wood. Using plastic-covered graph paper to measure the proper amount of filler and catalyst, Minick mixes the filler. After he packs the filler, he uses a knife to strike the repair flush with the surface, which will reduce sanding later. Areas of the leg that will be glued have been masked off. The filler cures quickly.

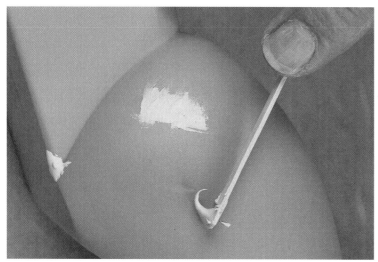

Glazing putty smooths out blemishes in primer—After Minick primes a leg, he sands it out to reveal tiny surface dents and nicks, which he fills with blue glazing putty. When that's dry, he'll sand, re-prime and then lightly sand again in preparation for paint.

Why new car finishes work on wood

Furnituremakers may question the wisdom of using automotive finishes on wood. After all, aren't car finishes brittle—meant for relatively immobile surfaces like metal instead of dimensionally unstable substrates like wood? Although that argument was true in the past, it is no longer accurate. Automotive primers, aerosol paints, clear-coat finishes and touch-up paints have changed because car components have changed. The latest materials, such as high-impact plastics and composites, are used to manufacture car bumpers, trim and door panels. So paint makers have had to reformulate their coatings to accommodate increasing flexibility. This flexibility allows woodworkers to use car-finishing products on wood, which is notoriously unstable. If you don't care to use finishes from the auto-parts store, you can use most general-purpose aerosol primers, paints and clear coatings to get equally stunning results.

assembly, it eliminates complicated masking and leads to better finish results.

When choosing stock for your project, think about which components will require special needs. For example, if you decide that certain parts must be real smooth, then maple, poplar and birch are good wood choices. However, if you want to show a bit of wood texture through the paint, then open-grained woods, such as oak and ash, are more appropriate. I wanted smooth, glossy black legs on my coffee table that would enhance the figured-mahogany veneer top and apron. In addition, I wanted the legs to be hard to guard against knocks. For these reasons, maple was the logical wood choice. But as far as the painting goes, the wood used is irrelevant, really, as long as you are careful with the under-paint treatments.

Preparing surfaces and equipment

The key to getting flawless painted furniture is meticulous surface preparation. The monochromatic nature of paint dramatically magnifies minor flaws that would otherwise go unnoticed under a clear finish. Small tearouts, hairline cracks in knots, stray sanding scratches and other seemingly minor defects must be filled and smoothed before painting. This may sound like lots of dismal work, but if you follow car surface-preparation steps, you can reduce the drudgery.

Sanding and filling

All parts should be thoroughly sanded to at least 180-grit and inspected under a strong light; then use auto-body fillers to level off any voids. These polyester fillers (familiar brands include Bondo and White Knight) work exceptionally well at repair because they tenaciously stick to raw wood, cure quickly, sand easily and accept most kinds of oil-based and latex primers and paint. Best of all, they don't shrink. On the down side, they smell bad and have a short working life once mixed, usually less than 15 minutes.

For the coffee-table legs, I filled dents and nicks with two-part auto-body filler (3M's 2K Lightweight putty). I even built out an edge that had been clipped off on the bandsaw (see the top photo on the facing page). I also filled in the knots. No matter how sound they look, knots always have cracks that show through the paint. Knots often contain resins, too, especially in softwoods. So once the filler in the knots had cured (about 30 minutes), I sanded them flush and spot-sealed the knots with shellac just to be safe. Finally, to make the edges of the medium-density fiberboard (MDF) top perfectly smooth, I used some spackle (see the story at right).

Setting up a makeshift spray booth

I don't have a paint booth in my home shop, so before I prime or paint, I set up a crude but effective painting area in my garage (see the photo on p. 172). Ventilation for my plastic spray booth is provided by a box fan that draws outside air through an open rear door and exhausts it through a partially opened garage door. I also use a good organic-vapor respirator to protect myself when I'm using aerosol cans to spray primer and high-solvent lacquers.

Priming and puttying

Primers serve the same functions for painted finishes as sealers do under clear coats. Primers seal in resins and extractives that may discolor the paint, provide a uniform non-porous base for the color coat and highlight any defects that were missed in the filling process. Aerosol primers are sensible to use if you're painting relatively small areas. I often use automotive high-build, scratch-filling primers under pigmented-lacquer paints. High-build primers are easy to apply, sand like a dream and fill in tiny nicks and pits in wood. Adhesion tests in my shop show that automotive primers are completely compatible with high-solvent lacquers, but marginally compatible with oil paints and not at all with latex.

When I buy primer at the store, I pick up several different brands of cans and shake each until I hear the little agitator ball dislodge. I pick the can that takes the longest for the ball to loosen because, generally, this means the primer contains a higher percentage of solids. Primers with more solids do a better job and are easier to sand.

Use spackle to fill voids in edges

The medium-density fiberboard (MDF) edges of my coffee-table top posed a unique finishing problem for me. Because the top was veneered, I needed a way to hide the MDF core. Edge-banding with solid wood was an option, but that didn't fit my design. I ruled out veneer as well because of the shaped edge that I wanted. So I decided to paint the edges black, like the legs. But first I had to prepare the surface of the MDF for primer.

MDF absorbs finish like a sponge, and the small pits in the core must be filled or they will show through the paint. A few finishers tackle this problem by using glazing coats; this technique requires real skill. Large furniture manufacturers solve the problem by spraying on two-part edge filler/surfacer, but it is expensive, hard to find and requires specialized spray equipment. I avoided all this by wiping a coat of wallboard spackling compound (made by DAP) on the exposed MDF edges (see the photo above). The spackle sands easily, fills the pits and provides a good base for the primer. To save yourself some work, mask off the top and bottom of the tabletop before you start spackling the edges.

Spackle fills voids in medium-density fiberboard—After masking off the veneered top of his table with paper and acrylic adhesive tape, the author rubs wallboard spackle onto the MDF edges. The spackle adheres well, dries quickly and sands beautifully.

Allow the primer to dry thoroughly (it should powder easily when sanded), and then inspect the piece carefully. You'll be surprised at the number of imperfections that will appear on your supposedly smooth wood. You must fill the tiny defects, or they'll show through the paint. Don't use the two-part auto filler this time, though, because it won't stick to the prime coat. Instead, use an automotive glazing putty, which is designed for application over primer (see the bottom photo on the facing page). 3M's Acryl-Blue

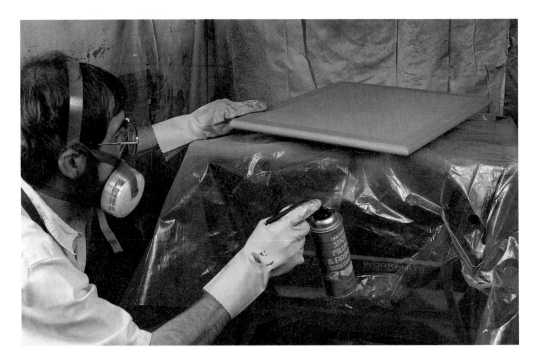

Simple tools improve the quality of a paint job—To prime the spackled MDF edges of an end table, Minick places it on a lazy Susan mounted to his bench. This allows him to spin the work as he sprays. Because the push buttons on aerosol cans are awkward, he uses an inexpensive plastic trigger handle.

Glazing putty suits my needs, but any non-shrinking brand should work.

Sand the primer and dried putty smooth. Next apply a final coat of primer. Then sand again—this time to at least 220-grit but not finer than 320-grit. While you're sanding, be careful not to cut through to the wood, or you will have to re-prime. The object of this final sanding is to level and smooth the surface but still leave some tiny scratches in the primer. This slight texture, called tooth, makes a better bond between the primer and topcoat.

Painting, clear coating and rubbing out

Aerosol paint cans are available in different colors, gloss levels and brands. I have had good luck using both Plasti-kote and Krylon on furniture. Aerosol paints that are low-gloss sand easier than high-gloss ones, but I prefer the high-gloss variety because their higher resin content adds to the durability of the finish. You shouldn't be overly concerned about the actual glossiness, however, because the final sheen of the project will be controlled by the clear coat.

To start painting, I mist a tack coat of paint over all the primed area. Then I spray several light coats to fill in the blanks until the entire surface is covered with a level wet coat. Continued painting at this point will result in runs or sags. Let the solvent evaporate for five minutes or so, and then lay on another coat the same way. Two or three coats are usually enough to provide sufficient color build on a well-primed substrate.

For the tabletop's edge, I overcoated the spackled and sanded edge with the same automotive primer and paint that I used for the legs. The only differences were that I masked off the top and then used a lazy Susan to hold the work (see the photo above).

The clear coat is the final touch that sets apart an average paint job from a real show-stopper. Clear coats not only protect the paint from occasional dings but also add depth to the finish, which is more suitable for fine furniture. In addition, clear coats unify components by providing a consistent sheen over the entire piece. And clear coats are easier to rub out and repair than paint.

For peace of mind, I usually choose my clear finish from the same resin family as the paint. I used an aerosol automotive clear acrylic on my table project, but any good clear lacquer will work. For the tabletop edges, I clear coated the paint with Pratt & Lambert #38, which is the same

For a high-gloss finish, rub out the clear coats with liquid automotive polishing compound. As a finishing touch, Minick buffs out the clear-lacquer topcoat (one of five) on a coffee-table leg.

varnish I used on the mahogany-veneered top and apron. Four or more clear coats may be needed to achieve a good film thickness (3 mils to 4 mils). Remember, some film will be lost when rubbing out, so compensate for this. Make sure that your paint is completely dry before you clear coat. I like to wait several days.

For rubbing out clear coats to a high luster, I like to use liquid automotive buffing compounds (not paste compounds). I've found that car buffing compounds are easier

to use than those carried by most wood-finishing-supply places. Both 3M and Meguiar's offer good compounds for polishing. Meguiar's has several formulas with different abrasive levels for hand-rubbing or power buffing. Let the clear coats dry a day or so, and then buff out to whatever sheen you desire (see the photo above).

REPAIRING A WORN FINISH WITHOUT REFINISHING

by Pinchas Wasserman

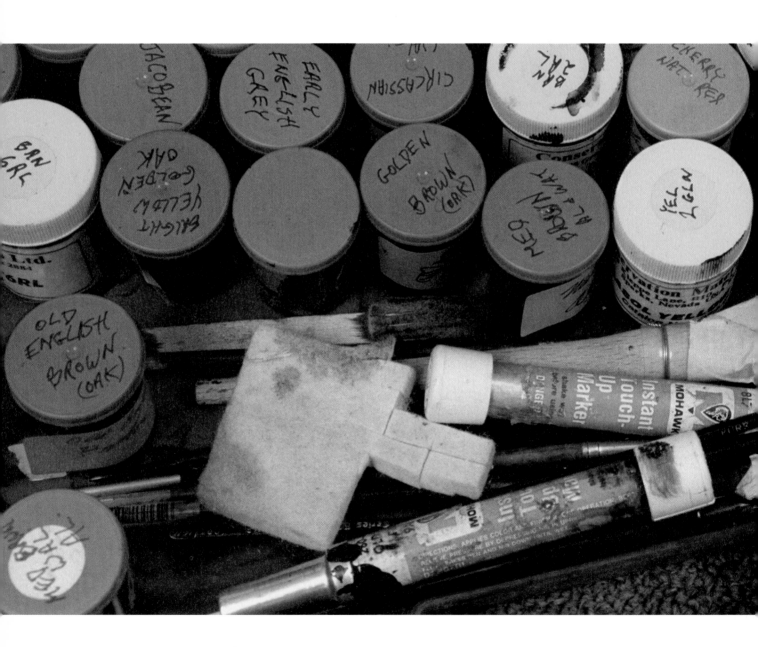

As a restorer, primarily of 20th-century furniture, one of my typical problems is how to improve an existing finish for a customer who is not ready to have the furniture stripped and refinished. More often than not, the furniture doesn't really need such drastic measures.

In cases like that, I've found alcohol-soluble dyes to be the most effective solution among the options available. These dyes receive mostly peripheral treatment in discussions about coloring wood. They are vastly more difficult to apply than oil-based pigment stains, and many of them are not as lightfast as water-soluble anilines. Yet when it comes to touching up existing finishes, I

regard alcohol-soluble dyes as the premier colorant. Their capacity to bite into a finish or sealed wood makes them uniquely suited for restoring worn finishes (see the top photo on p. 180).

Only your chemist knows for sure

Two kinds of alcohol-soluble dyes are commonly available: basic dyes and metal-complex dyes. Both may be sold as aniline dyes (see the box below). Basic dyes, available through many woodworking catalogs, are the most common and are available in a greater range of premixed wood-tone colors. Not all of these are considered lightfast. But with small areas of worn or chipped finishes, lightfastness is not that important.

Metal-complex dyes, which are manufactured by Ciba-Geigy and BASF (see Sources of Supply on p. 180), are less common, more expensive (not that you'll need much for touch-up work) and more resistant to fading. For practical purposes when touching up finishes, there is not a great difference between the two. Both are excellent, powerful dyes.

Mix dyes with alcohol and shellac, and apply small amounts with a brush or cloth

After mixing dyes with denatured alcohol, I combine the solution with a finish resin. I prefer shellac because it is less toxic and eas-

What's in a name?
The story behind aniline dyes

by Jeff Jewitt

In 1856, a young English chemist named William Perkins was trying to find a way to make synthetic quinine. He mixed acidified potassium dichromate with aniline, a chemical derived from coal tar. The bluish-colored precipitate was obviously not quinine, but Perkins

had the foresight to recognize that this product had potential as a dye. He set up a factory to manufacture the stuff and ushered in the modern era of dye making.

To distinguish the new dyes from the older, natural dyes that were still widely used, the terms aniline dye and coal-tar color

were applied to these products. Back then, the principal ingredient in most dyes was aniline. Although, aniline may or may not be used in the process today, the term aniline dye has stuck and is used loosely to refer to the entire class of synthetic dyes.

No stripping required. Working with alcohol-soluble dyes and a fine brush, the author makes repairs to this walnut desk that will be virtually undetectable.

Padding lacquer applied by cloth— Applied with a quick, buffing motion, padding lacquer blends finish repairs and seals in alcohol-soluble dyes.

Squirrel-hair brushes for blending large areas of color—Keep brushes soft and supple with occasional dips in denatured alcohol.

ier to manipulate than lacquer. If I make a mistake when applying the finish, it's relatively easy to remove with alcohol, provided the dyes are used on top of the finish and not on raw wood. The denatured alcohol in the finish may damage the surrounding surface, but that is easily repaired by applying padding lacquer and rapidly buffing the surface with a lint-free rag (see the center photo at left). Alcohol-soluble dyes also can be used to tint lacquer, which is a good choice if the repair area is large. I've had good luck with a brushing lacquer such as Deft's clear gloss. It dries relatively slowly.

Typically, I often use less resin for the initial coloring, then topcoat with a greater proportion of resin. First I dissolve the dye in pure alcohol, and then I add shellac in small amounts. I use mostly super blond shellac that I mix from dry flakes. It seems to work on both light and dark finishes. Zinsser's premixed clear shellac (available in most hardware stores) is a less expensive substitute, and its water and wax content is not a factor in touch-up work. Buttonlac, less refined than orange or blond shellac, is good for dark finishes and adds a little opacity to a dye. Alcohol dyes are transparent. If you need true opacity in a stain, you must turn to pigment powders, Japan colors or glazing stains to do the job.

I apply alcohol dyes in one of four ways: with a brush, a padding cloth, felt or an air-brush. Pointed red sable brushes, no. 2 and no. 4, are my most-used brushes for fine detail work. For larger areas, I use squirrel-hair polisher's mops, no. 4 and no. 8, the smaller being the more useful (see the bottom photo at left).

To match an existing finish, orange and blue-black dyes will suffice to create many of the common furniture browns. The steady addition of small amounts of black will lead you through maple browns to walnut. Often, the addition of yellow or red will swing the color one way or another. Try out your dye and shellac mix on a small area, and topcoat it to see how it will look. The topcoat will make the color look bolder and darker.

BURNING IN INVISIBLE REPAIRS

by Robert Judd

Blending several colors on a hot knife yields just the right shade for an invisible repair. After mixing the resin thoroughly with a small screwdriver, the author flows the molten resin into the damaged area. Burn-in repairs will accept virtually any topcoat.

Whether trying to rescue a hand-polished antique or save a chunk of exotic hardwood from the scrap pile, burn-in sticks and a hot knife can make virtually invisible repairs. A distant cousin of sealing wax, burn-in techniques have been around for centuries. Burning-in can disguise or cosmetically cover scratches, chips and other damage in a finely finished piece. The repair sticks are available in a host of colors and tints and readily intermix to match any color. Because the various resins and materials used to formulate the repair sticks accept virtually any topcoat, the repair's finish can precisely match the original. A burn-in repair is a reversible process; you can scrape it out and start over if you don't like the color match or result. These simple repairs can save hours of rebuilding or refinishing time.

In this article, I'll look at the burn-in repair process: color matching, preparation, filling the defect, leveling the repair, adding wood-grain details and touch-up finishing. Before trying to repair your favorite piece of furniture, I suggest practicing on a scrap panel of finished wood. It's a simple process, and the basic skills can be learned in a couple of hours. But you also can spend a lifetime perfecting those skills. If you're dealing with a valuable antique, you should consult a professional prior to attempting any repair.

Tools and materials

Burn-in repairs don't require a huge inventory of expensive tools, compressors and spray guns. A hot knife, a selection of shellac-resin sticks, a few other miscellaneous supplies and a can of spray lacquer can get you through most of the repairs that you'll encounter (see Sources of Supply on p. 188).

The sticks are melted into place using a hot knife, either a thermostatically controlled electric knife or a manual knife with an alcohol lamp using denatured alcohol as its fuel. Don't try to use a candle as a heat source because it produces soot that mixes with and changes the color of the repair material. I've also heard of using soldering irons, but I don't advocate the practice. The iron is too hot and can burn the stick shellac, as well as the wood you are trying to repair.

I prefer the electric knife because it poses little fire hazard, and it maintains a consistent temperature. My tool of choice is a lightweight, slim and highly maneuverable knife made by Hot Tools, Inc. (see Sources of Supply on p. 188). But I also carry an alcohol lamp in my tool kit for those occasions when no electricity or lack of room makes a manual knife a logical choice.

The shellac-resin sticks are available in a wide variety of colors and tints. And different colors can be melted and mixed together on the hot knife to make the exact shade necessary for a perfect match. Some colors are available translucent or opaque, which can help make certain repairs far less noticeable. The translucent shades are perfect for repairs when just the finish is damaged and not the underlying wood. This situation is identified by no telltale color changes in the damaged area. A color change indicates broken wood fibers that will require opaque colors. But opaque and translucent colors can be intermixed to suit any repair situation.

Other than the new Concept 2000 repair system by Mohawk (see Sources of Supply on p. 188 and the box on the facing page), shellac resins are very brittle when cool and can chip when used on corners or other areas subject to sharp impact. The resin has no strength to hold nails or screws, so it should be used for only cosmetic repairs.

Color-matching

Probably the most important step in-volved in making a good repair is color matching. It is an art that takes some practice, but color-matching can be learned. Most wood has a background shade that is dominant in creating the impression of color. Quickly glance at a piece that needs repair. The color you see in that quick look is the one that you want to match. Now look closely at the piece. You will imme-diately notice that it is made up of many colors that are subtly blended. By carefully selecting the right shades of resin sticks, repairs also can be subtly blended.

As a general rule on medium and dark finishes, if you can't get an exact match, it is better to make your repair darker than the surrounding wood. With lighter woods, it is better to err on the lighter side. This will

Burning-in for beginners

As I was putting together the information on burn-in repairs, my Mohawk finishing products salesman introduced me to Concept 2000, a new repair system that was recently introduced. After working with Concept 2000, I found the system eliminates or minimizes many of the common problem areas, especially for the novice. The repair sticks are flexible, so repairs made with the new sticks won't be as fragile as the brittle traditional sticks. And the halo effect, a glossy ring left by traditional burn-in repairs, is gone because the sticks dry flat, making it easier to blend in the repairs.

Lower temperatures for better results: What really sets the Concept 2000 system apart from the traditional method is the fact that the burn-in knife is cooler than the old-style knife and need never touch the surface of the wood. Because the hot knife is kept away from the finish, there is less chance of creating more damage when making the repair. Mohawk offers a special, temperature-controlled knife for melting the burn-in sticks. However, I modified my standard burn-in knife by adding a rotary, lamp-dimmer switch for controlling the temperature. Too much heat causes the melted sticks to boil, which leaves air bubbles in the repair and darkens the color.

Repairs are made in the traditional manner of flowing melted material into the damaged area, leaving a dome-like mass. Air bubbles are removed by allowing the repair to cool for 10 to 15 seconds and then pressing on the material with a fingertip dipped in Plane Balm, Mohawk's skin and finish protector. With the easy melting, flexible repair sticks, I found this step to be rarely necessary.

Leveling the repair: To level the repair with the surrounding wood, Mohawk includes a patented device called an leroplane, which takes the place of a sharp chisel and allows even a novice to pare away excess material without fear of gouging the surface being repaired. The leroplane is a large round blade mounted in a non-marking plastic frame, as shown in the photo below. The height of the blade relative to the repaired surface can be adjusted with a top-mounted knurled wheel. The blade is adjusted to just skim the top off the mounded repair, which should be allowed to cool four to five minutes. The leroplane is pushed through the repair material with a slight rolling or twisting motion. Keep adjusting the blade to take off paper thin slices until the repair is almost dead even with the surface.

If necessary, further leveling can be accomplished with Mohawk's Level Aid, which is used as naphtha would be for a conventional repair. Level Aid is applied with a felt pad, a rag or 600-grit, wet-or-dry sandpaper. Limit the sanding and smoothing to just the repair area with as little overlap onto the surrounding finish as possible.

Graining can be done with appropriately colored, fine-line artists' markers in the normal manner. Spraying lightly misted coats of lacquer until the depth and gloss of the repair's finish matches the original provides a quality repair in a minimal amount of time.

Trimming burn-in repair material flush is done without fear of damaging the surrounding finish, thanks to the leroplane. The leroplane is included in Mohawk's new Concept 2000 repair system, which makes burn-in repairs easier for the beginner.

Placing shellac-resin sticks directly on the piece to be repaired is the best way to get a sense of the color match. This technique also helps in selecting the shades that need to be blended, such as the three different colored sticks shown here.

help prevent high contrast situations, which stand out and call attention to your repairs.

A trick that helps me match colors is to place some sticks on the piece to be repaired, as shown in the photo above. Looking at the color in situ is the best way of knowing if a single stick will make a good match.

Because there are so many formulas and proprietary resins used, the manufacturers' color names can be a bit confusing. Each manufacturer has its own idea of what color "antique mahogany" or "golden oak" should be. The key is to match the wood to the repair stick color, not the name. It's all right to use a dark walnut to repair a piece of mahogany if the color match is good.

Often you will need to mix or blend colors on the knife while the resin is in a liquid state, as shown in the photo on p. 181. As a rule of thumb, if you are mixing more than three colors, you are probably in trou-

ble; just wipe the knife clean on a rag and start again.

By purchasing a set of sticks, you will have a better chance of having the right colors on hand. I recommend buying a set of small, 2-in.-long sticks because you can get a great selection for a modest price. Even though 2 in. is the smallest size, the sticks will still go a long way because so little material is used for each repair.

Making the repair

Before starting a repair, the first step is to clean the damaged area using a rag dampened with VM&P (varnish makers and painters) naphtha. This solvent will remove wax and dry in a matter of minutes, leaving a powdery residue that can be wiped off with a clean cloth. I then apply Burn-In Balm (Mohawk) or Patch Lube (Star) around the damaged areas. These thick ointments, the consistency of Vaseline petroleum jelly, act as heat sinks and help

Smooth the repair with the hot knife by drawing the knife over the resin and then wiping the excess resin from the knife. Keep the knife moving because it is easy to burn the surrounding finish.

prevent burning the surrounding finish. They also act as repellents, so the resin adheres only in the damaged areas. A small bottle lasts a long, long time and makes repairs easier and neater.

Before beginning the repair, clean the knife to avoid any color contamination from residue clinging to the blade. Preheat an electric knife three to five minutes or a manual knife 30 seconds, and wipe it clean with a rag or scrape it clean with a single-edge razor blade. When you apply the knife to the stick, the resin should flow smoothly rather than bubble or boil, a clear sign the knife's too hot. After melting enough material to slightly overfill the damage, allow the molten resin to flow into the damaged area, creating a slight dome of excess material.

Wipe the knife clean, and then smooth the repair by drawing the knife over the resin, always working toward you, as shown in the photo above. Repeat the process of wiping the knife clean and stroking the repair at least three times or until most of the excess material has been removed. But don't try to get the fill dead even with the knife because it's easy to damage the surrounding finish, especially if you don't keep the knife moving.

Leveling the repair

When finished with the hot knife, the repair should be slightly proud of the finish. As the resin cools, it hardens almost immediately. I shave off the remaining excess filler with a chisel I hone to a razor's edge on an 8,000-grit Japanese waterstone.

After carefully cutting back the repair, you may find bubbles or unevenness. One of the characteristics of a burn-in stick that

Buffing the repair with solvent and a felt pad softens the resin, levels it with the surrounding wood and leaves a smooth surface. Excessive resin should first be pared away with a razor-sharp chisel.

makes it such a good repair material is its forgiving nature. Just reapply heat, and the resin will liquefy again whether it's been 30 seconds or 30 years since the repair was made. This reversibility means the repair can be pulled out if you botch the job or if a better technique is developed.

The next step in the leveling process is to wipe the resin with a felt pad and a solvent to help smooth the repair, as shown in the photo above. I like to use Brasiv (Mohawk) or VM&P naphtha. Often I use cigarette lighter fluid from Walgreen's. The lighter fluid is essentially naphtha, but the little flip-top can is a very handy dispenser, which keeps spills to a minimum.

Although they will rarely damage any cured finish at least 30 days old, solvents should be tested on an inconspicuous spot. Brasiv is fast acting, so it requires some care. Using a clean felt pad slightly moistened with solvent, I rapidly rub the repair. The solvent softens the resin, and the felt absorbs any excess material while polishing the surface smooth.

Many people use 400-grit wet-or-dry sandpaper to smooth repairs. This technique requires a very gentle touch and often creates more work if you damage the finish around the repair. I try not to use sandpaper and especially don't recommend it for beginners. Things can get out of control in a hurry, particularly on more modern pieces that have toner in the finish instead of being stained in the more traditional manner. Sanding, even with 400-grit paper, will cut

through the toned finish in short order, and it is difficult to patch unless you are skilled with an airbrush. If you must use abrasives, keep in mind that a finish is not very thick, and you can sand through it in less time than it takes to talk about it. Try starting with less aggressive 600-grit paper, and use a very light touch.

Adding wood-grain details

At this stage, the repair should be flat and the color close to the shade of the background wood, but it still won't look quite right. The missing ingredient is the natural grain lines of the wood. The grain lines need to be duplicated to commit what the French call trompe l'oeil, literally, to "trick the eye." Wood-grain lines of the appropriate size and color are drawn with fine-line graining pens or painted with a brush and pigments blended with lacquer and a little retarder, as shown in the photo above. Artists' pens and pigments also can be used and offer an even greater selection of colors.

For best results, be sure to follow the natural grain lines. Run the pen lines past the edges of the repair, and avoid hard edges. Smooth in coloring pens and brushed colors with your finger to blur the lines. These grain lines are vital to an invisible repair, so take the time needed to get them right.

Getting the finish right

Though the repair is now smooth, grained and color-matched, it will probably still stand out. The sticks leave glossy repairs, so the sheen of the repair won't match the surrounding wood's sheen. The location of the repair and the finish of the piece determine the next step.

In most situations, the gloss of the repair can be adjusted with either Burn-In Seal (Mohawk) or Sheen (Star) aerosol lacquers. These aerosol lacquers are specifically formulated to reduce the gloss of the burn-in repair to match most wood finishes. I am not sure how they work, but they sure do the job. I lightly mist on several coats of the lacquer until the repair blends in with the rest of the finish. If necessary, you can topcoat with the original finish for a perfect match.

This technique works fine in most situations. One problem area, however, is tabletops, particularly those with a high-gloss finish. A high-gloss finish works like a mirror, reflecting light and accentuating scratches, nicks and even repairs.

Repairs to these finishes are a compromise, and no technique yields perfect results. With that idea clearly in mind, I like to use MicroMesh Abrasives, available from C.W. Crossen (see Sources of Supply above), to reproduce the luster of a high-gloss finish. These super-fine abrasives, available in grits as fine as optical grade (12,000-grit) are worked with a drop or two of water from the finest grits downward toward the coarser grades, which is the reverse of the usual procedure, until a reasonable duplication of the surrounding sheen is achieved.

WHICH FINISHES ARE FOOD SAFE?

by Jonathan Binzen

I was hoping to compile a list of foolproof products and strategies for food-safe finishing. But I soon discovered that it wasn't going to be that easy. What I found, after scores of conversations with chemists and regulatory agencies, finish manufacturers, finishing experts and woodworkers, is that although there are a few finishes that everyone agrees are food safe (see the box on p. 193), those finishes tend to be the least protective. I also found that the great majority of finishes are in a kind of limbo, with many experts saying most are fine for use with food but with others saying they should be avoided because there are some lingering questions about their safety.

Coatings you can cut on

For cutting boards and the like, you can cut the confusion in half. Wood finishes can be divided into two broad categories: film-forming finishes, which harden in a thin layer on top of the wood, and penetrating finishes, which harden (if they do harden) in the wood rather than on it. When choosing a finish for a cutting or chopping surface, you can start by ruling out the film finishes. Although film finishes like polyurethane, nitrocellulose lacquer, varnishes and epoxy form a hard surface and are considered by many to be nontoxic when cured, they aren't impervious to knives. By cutting on boards with these finishes, you'll eventually slice through the film, inviting water underneath and compromising the finish.

That leaves you with the penetrating finishes to choose from. From the standpoint of food safety, this group can be chopped in two as well. On one side are what I'll call unmixed oils—pure tung oil, raw linseed oil, mineral oil and cooking oils such as walnut oil. These unmixed oils are all naturally occurring substances, are all sold in their pure form and are all perfectly edible (although not all delectable). On the other side are what I'll call mixed oils—boiled linseed oil and the range of oil-and-varnish mixtures often sold as teak oil, tung oil finish and Danish oil. The mixed oils are synthesized blends of oil, resins, driers and other ingredients whose identity often won't be revealed on the can.

Finishes you invite to dinner. For protecting cutting boards and other abused and oft-washed items, penetrating oils are best.

Mixed oils

As a class, the mixed oils offer considerably more protection from moisture and staining than the unmixed oils because of the resins and other additives most mixed oils contain. (Boiled linseed oil is an exception. It does not contain resins and is not as water-resistant as the other mixed oils.) Mixed oils are made easier to use by the addition of driers. The metallic driers in a tung-oil based varnish, for example, make it easier to work and quicker to cure than pure tung oil.

But the safety of mixed oils as a finish for cutting boards and other items in contact with food is the subject of debate. It is the heavy-metal driers, primarily, that cause some people to consider mixed oils unsafe for food surfaces. The metals used no longer include lead or mercury, but one chemist told me that the currently used driers aren't above suspicion and that he wouldn't use a finish with heavy-metal driers on a cutting board. "Twenty years ago, nobody was worried about lead. Look what happened," he said.

By far the majority of people I spoke with, however, consider that mixed oils, once fully cured (a process that can take up to a month for some finishes), are probably fine for contact with food. Watco Danish oil, for example, has been used for years as a finish for cutting boards and bowls. According to one chemist I spoke with at Flecto Coatings, which now manufactures Watco, the company has never had any complaints. He said he thinks the question of food safety is "substantially a nonissue."

But Watco isn't pitched as a food-safe finish, and the company won't recommend it for such use because, as I was told, "We haven't done the testing." The testing he was referring to is a lengthy and expensive process of assessment conducted by the Food and Drug Administration (FDA). Watco oil hasn't been found unsafe, of course, but unless it is tested and found to comply with FDA regulations, the company won't accept the legal liability of proclaiming it a food-safe finish.

One company that has done the testing is Behlen's, makers of Salad Bowl Finish. They have been touting this mixed oil (probably tung-oil based, but they won't tell) for years as food safe, but only recently had it tested by the FDA (it passed). The process was costly as well as protracted, and Behlen's had to pull the finish from distribution during the testing. According to several finishing experts, passing the FDA tests doesn't necessarily make a particular finish different from others in its class; it has been tested, and the others haven't. All of which means that the consumer is left without a definitive answer on whether most mixed oils are food safe.

Unmixed oils

With unmixed oils, there is no such dilemma. We know that pure tung oil and raw linseed oil are edible. And you can lean back and drink walnut oil or mineral oil right out of the bottle, if you want to.

Pure tung oil will provide a much better moisture barrier than any of the other unmixed oils (and some of the mixed oils), but it does present difficulties in application, requiring multiple coats with a day or so to dry between coats and at least a week of final curing time. Pure tung oil is an ingredient in a slew of other finishes, many of which use tung oil in their name. If you want the unmixed version, look for "pure tung oil" or "100% tung oil" on the label.

Applying raw linseed oil is also a long-term project. Without the driers that are added to make it boiled linseed oil, raw linseed oil can take several weeks to cure. Even then, it doesn't provide good water-resistance and will have to be reapplied fairly often.

Walnut oil is probably the best of the cooking oils for use as a finish, because, unlike olive or peanut oil, walnut is a drying oil—it polymerizes within a few days of application, its molecules linking together so it becomes inert and cannot go rancid. There's no trick to applying walnut oil—just soak the surface, triple-soak the end grain, let it sink in and wipe off the excess. It won't provide much water-resistance and will need to be reapplied frequently.

Food fight:
Wood vs. plastic cutting boards

Cutting controversy. Among researchers, plastic cutting boards were once favored as more sanitary, but wooden boards are gaining proponents.

The debate over whether plastic or wood is the better material for cutting boards rages on. There is research on both sides of the issue. One prominent study shows that wooden cutting boards are less likely than plastic ones to harbor bacteria after being used to cut raw meat. Other studies show just the opposite. The Food and Drug Administration (FDA) and other agencies that oversee the food-service industry have backed off their previous stand that only plastic should be used. The FDA now recommends that boards be made of plastic or of "hard maple or equivalently hard, close-grained wood." The FDA stresses that to make cutting boards easy to clean, it is important to use materials "free of cracks and crevices," and to "avoid cutting boards made of soft, porous materials."

With any cutting board, the primary danger (assuming you wield your knives carefully) is from food poisoning. This can occur when a food that will be eaten uncooked is chopped on a board that has previously been used to cut raw meat or raw fish. To prevent such cross-contamination, the FDA recommends that after cutting raw meat or fish, you should wash the cutting board with hot water, soap and a scrub brush. Periodically, sanitize the board with a solution of chlorine bleach (1 tsp. bleach to 1 qt. water). Flood the surface with bleach, let it sit a minute, then rinse it off. You can protect yourself further by using one cutting board exclusively for foods that will be cooked and another one for foods that are ready to eat.

Slicing to the heart of the matter. With either wood or plastic cutting boards, segregating raw meats from ready-to-eat foods is the way to avoid contamination.

If you ask professional makers of cutting boards what finish they use, whether in one-man shops or at major manufacturers, it's almost certain you'll find they use old-fashioned mineral oil. Although it's a derivative of petroleum, mineral oil is odorless, tasteless and colorless, completely inert and approved as a food additive by the FDA. It is the same stuff you see in the drug store sold as a laxative. Some makers I spoke with use it straight, others blend it with beeswax or paraffin, and everyone had a twist on how

it ought to be applied. (For information on making and applying a mineral oil and beeswax finish, see the box on p 194.)

Many makers use mineral oil because of the convenience and low cost. It certainly isn't a miracle finish. It never dries, and to maintain even a modest amount of protection for the wood, you need to reapply it often. But it is extremely inexpensive, and because the stuff is completely inert, it will last indefinitely in the bottle. And it ranks with the least fussy of all finishes. As with walnut oil, you simply drench the cutting board, let it soak and wipe it dry. Various companies sell mineral oil explicitly as a wood finish, but such products are typically twice the price of mineral oil sold as a laxative in the drug store. The drug store variety may be slightly more viscous, but it is the same substance. If you want it a little runnier to make it penetrate the wood more deeply, you can heat it gently on the stove.

Which finish for woodenware?

The considerations for finishing wooden bowls, serving and cooking implements, plates and trays are the same as those for finishing cutting surfaces, with a few exceptions. For stirring spoons, pasta forks, spatulas and other implements that will see duty in bubbling liquids or sizzling solids, some kind of penetrating finish would be preferable to a film finish. Although you won't be cutting on these finishes, the combination of heat and water will eventually undermine even a rock-hard finish like epoxy. And certainly any duty in the dishwasher (not a great idea for wood with any finish) will eventually whiten and crack a film finish.

One place where a film-forming finish would be appropriate is on items like trays that won't be subjected to high heat or constant washing but would benefit from a water-repellent finish. Chris Minick, a chemist and contributing editor to *Fine Woodworking*, suggests that if you are

Edible finishes

Pure tung oil. Extracted from the nut of the china wood tree. Used as a base in many blended finishes. Available from catalogs and hardware stores. Difficult to apply, requires many coats, good water-resistance.

Raw linseed oil. Pressed from flax seeds. Not to be confused with boiled linseed, which contains metallic driers. Listed as a food additive by the Food and Drug Administration (FDA). Very long curing time, good looks, low water-resistance, frequent reapplication.

Mineral oil. Although derived from petroleum, it is colorless, odorless, tasteless and entirely inert. Sold as a laxative in drug stores and as a wood finish in hardware and kitchen-supply stores. Simple to apply, low water-resistance, frequent reapplication.

Walnut oil. Pressed from the nuts of the walnut tree. Sold as a salad oil in health food stores and in large grocery stores. Walnut oil dries and won't go rancid. Easy to apply, frequent reapplication.

Beeswax. The work of the honey bee. Can be mixed with an oil to create a better-smelling, slightly more water-repellent finish. Sold in woodworking and turning catalogs.

Carnauba wax. Derived from the Brazilian palm tree. Harder than beeswax and more water-resistant. Can be used straight on woodenware as a light protective coating or a topcoat polish. Sold in woodworking and turning catalogs.

In the welter of opinions about which finishes are food safe and which are not, a few naturally derived, unblended, no-hidden-ingredients, certainly nontoxic finishes stand out.

Shellac. A secretion from the lac bug. Harvested in India. Super blond shellac in flake form is the most water-resistant variety. A film-forming finish. Sold in woodworking catalogs and hardware and art supply stores.

Nothing. Available everywhere. Makes a reasonable finish for woodenware. No application time. Free.

A recipe for one sweet finish

The food-safe finish that appeals most to me is one recommended by Jim and Jean Lakiotes, West Virginia makers of spoons and other kitchen items, as well as furniture. Their finish is a mixture of mineral oil and beeswax. To make it, warm the mineral oil in a saucepan over low heat, and melt a chunk of beeswax in it equal to about one-fifth or one-sixth the volume of the oil. (At high heat, there's a potential for fire. Be sure to keep the heat low, and consider using a double boiler.) As the wax begins to flake apart and dissolve, stir frequently. When the mixture is blended, pour it into a jar to cool and solidify. To apply, wipe on an excess of the soft paste, let it dry a bit, then wipe it off. If you want to apply it as a liquid, you can reheat it. Like any mineral oil or wax finish that will take a lot of abuse, this one will need to be reapplied often to afford decent moisture protection. But applying this fragrant finish is such a pleasure that you may find yourself looking forward to the task.

uncomfortable using a synthetic film-former, you might try shellac. If the shellac is selected and applied with care, it will provide a hard finish that can be sponged off and ingested with impunity. Minick recommends using only super blond shellac and buying it in dry flake form. Super blond, also known as dewaxed shellac, is ordinary shellac that has been refined to remove the wax that makes other grades of shellac less water-resistant. Buy it in flakes rather than pre-mixed, and use it immediately, because once it is in liquid form, shellac gradually begins to lose its water-repellent properties. But remember, even a drop of vodka from a martini or a dribble of wine will dissolve shellac and mar the finish.

A final option for finishing wooden-ware—an option that neatly skirts all these difficult issues—is to use no finish at all. If you don't mind chancing some stains and are willing to forego the luster an oil finish can add, it is a viable solution. With the right wood, something hard and tight-grained and ring-diffuse like maple, beech or cherry, there shouldn't be any major problems with going buck naked.

A CASE AGAINST FINISHING

by Pat Warner

It always bothers me when I begin applying the finish on a piece of furniture and suddenly realize I'm only halfway to completing the job. I work like crazy to apply good design, milling and joinery to the furniture I make. That should be enough. Now just flood with Danish oil and deliver. Right? Well, perhaps. Danish oil is an easy, cheap and often acceptable finish, but for furniture that will take a beating or for high-end work, a hard finish and some filling and coloring is often required. To obtain such a finish takes special skills, techniques and equipment and often large amounts of time and money. This is not woodworking. It's chemistry, abrasives, coloring, compressors, spray guns, resins, solvents, clean rooms and rubber gloves. And I'd rather not get tangled up in all of that if I can avoid it.

Finishes have their advantages, I admit. But when neither the environment nor the users are particularly threatening, a bare wood cabinet can be a refreshing change. Unfinished furniture is warmer both to the touch and the eye. It develops a nice patina and won't wear out a minute sooner than work that's French polished or sprayed with automotive acrylic urethane. If it does suffer an occasional insulting hand smear or wet glass mark, a simple sanding or steel wool buff-up will quickly restore the original look. Try that with a catalyzed lacquer or an acrylic.

When you finish wood, you emphasize the grain, color and figure, and this will limit its use in some applications. The soft, non-reflecting surfaces of unfinished wood, no matter the tree, play down the characteristics of the wood and put the material more in the service of the design.

A "no finish" finish is a natural with light woods like birch, beech or maple that will

Complete but unfinished—Fed up with finishing, the author never flowed finish onto his credenza. Two years later, the maple and yellow satinwood have taken on the subtler tones time gives to bare wood.

yellow badly under finish. These are beautiful woods that shouldn't be discarded for this idiosyncrasy. Left unfinished, these woods yellow a little, but with the advance of the patina, the color mellows, bringing up light tans and other tonal subtleties, as you can see in the photo of the sliding door of my credenza.

If you're hesitant about making an unfinished piece for the house or a client, make something for the shop: perhaps a jig, fixture or bench. Get some first-hand experience with bare stock, and see how it wears and ages. If you like it, think of how much more quality time you can invest in the next piece—time that would have been spent sanding, priming, sealing and rubbing out that finish.

ABOUT THE AUTHORS

Jonathan Binzen, formerly of *Home Furniture* magazine, now works as a senior editor of *Fine Woodworking* magazine. He learned woodworking in various shops in Philadephia, and then went to live for several years in Southeast Asia, writing, designing, and building furniture in Malaysia and Indonesia. He now lives in Newtown, Conn.

Andy Charron has been a professional woodworker since 1989 and is the owner of Charron Wood Products. He writes about woodworking and finishing for several publications, including *Fine Woodworking*. He is the author of *Spray Finishing* and *Water-Based Finishes*.

David Colglazier and his wife, Laurie, own and operate Original Woodworks, an antique furniture and trunk-restoration company in Stillwater, Minnesota.

Sven Hanson is a woodworker and professional carpenter in Albuquerque, N.M.

Dave Hughes is a professional finisher in Los Osos, Calif.

Jeff Jewitt is a small buisness owner, writer, and wood finisher and restorer. He owns and runs Homestead Finishing Products and J. B. Jewitt Co., Inc., which specializes in restoration and conservation of period furniture. Jeff is also the author of *Hand Applied Finishes* and two related videos. His latest book, *Finishing Step-by-Step*, will be published by The Taunton Press in early 2000.

Roland Johnson restores antiques and builds reproduction furniture and architectural millwork in his shop in St. Cloud, Minn.

Robert Judd is a professional furniture repairer and refinisher in Canton, Mass.

Kirt Kirkpatrick lives in Albuquerque, N.M. Formerly a journeyman patternmaker and boat builder, he now carves and builds furniture and doors.

Chris Minick is a finishing chemist and a woodworker in Stillwater, Minn. He is a regular contributor to *Fine Woodworking* magazine.

Frank Pollaro designs and builds custom furniture in East Orange, N.J.

Kevin Rodel designs and builds furniture with his wife, Susan Mack, in Powal, Maine. They have been building furniture, primarily in the Arts-and-Crafts tradition, for 11 years.

Mario Rodriguez is a contributing editor to *Fine Woodworking* magazine and woodworker living in Haddonfield, N.J. He teaches toolmaking, furnituremaking, and antique restoration at the Fashion Institute of Design in New York City. He is the author of *Traditional Woodwork*.

Gary Straub has been building (and sanding) furniture in Columbia, Mo., for 20 years.

Pat Warner is a woodwoker deep into routing, an instructor at Palomar College in San Marcos, Calif., and an occasional consultant for the router and bit manufacturing industry. He has two books published, *Getting the Very Best from Your Router* and *The Router Joinery Handbook*. He lives in Escondido, Calif. He also has a webpage at www.patwarner.com.

Pinchas Wasserman often travels to client's homes to restore furniture. He lives in Lakewood, N.J.

Tom Wisshack has been a woodworker for 35 years. He has plied his trade as a restorer in England, Germany, and the United States. He has published articles in *Fine Woodworking, Popular Woodworking, Woodwork,* and *Professional Refinishing* magazines. He lives in Galesburg, Ill., with his dalmation, Betsy.

Nick Yinger is a professional land surveyor in Kirkland, Wash.

CREDITS

Jonathan Binzen (photographer): 122, 124, 125, 193, 194

Anatole Burkin (photographer): 138, 139, 140, 142, 143, 144, 145, 146, 148, 150, 151, 153

Christopher Clapp (illustrator): 120

William Duckworth: (photographer): 21, 22, 23, 24, 25, 61, 62, 63, 65, 156, 157, 158, 159, 160, 162, 164, 165, 166, 178, 180

Aimé Fraser (photographer): 33, 34, 35, 36, 37

Michael Gellatly (illustrator): 70, 72, 90

Kirk Gittings (photographer): 80, 81

Dennis Griggs (photographer): 56

Boyd Hagan (photographer): 69, 70, 71, 73, 134, 135

Susan Kahn (photographer): 4, 6, 7, 8, 10, 29

Heather Lambert (illustrator): 53, 127

Vincent Laurence (photographer): 38, 39, 40, 57, 58, 74, 75, 76, 77, 78, 91, 92, 93, 94, 96, 97, 98, 99, 100, 101, 167, 168, 169, 170, 171

Robert Marsala (photographer): 18

Sandor Nagyszalanczy (photographer): 28, 30, 31, 32

Scott Phillips (photographer): 59, 60, 104, 189, 190, 192

Charley Robinson (photographer): 82, 83, 85, 86, 88, 89

Alec Waters (photographer): 12, 13, 14, 15, 16, 17, 18, 19, 20, 50, 51, 54, 55, 102, 104, 105, 106, 117, 118, 119, 120, 121, 128, 129, 130, 131, 133, 135, 172, 173, 174, 175, 176, 177

Matthew Wells (illustrator): 106, 181, 183, 184, 185, 186, 187

EQUIVALENCE CHART

Inches	Centimeters	Millimeters	Inches	Centimeters	Millimeters
$^1/_8$	0.3	3	12	30.5	305
$^1/_4$	0.6	6	13	33.0	330
$^3/_8$	1.0	10	14	35.6	356
$^1/_2$	1.3	13	15	38.1	381
$^5/_8$	1.6	16	16	40.6	406
$^3/_4$	1.9	19	17	43.2	432
$^7/_8$	2.2	22	18	45.7	457
1	2.5	25	19	48.3	483
$1^1/_4$	3.2	32	20	50.8	508
$1^1/_2$	3.8	38	21	53.3	533
$1^3/_4$	4.4	44	22	55.9	559
2	5.1	51	23	58.4	584
$2^1/_2$	6.4	64	24	61.0	610
3	7.6	76	25	63.5	635
$3^1/_2$	8.9	89	26	66.0	660
4	10.2	102	27	68.6	686
$4^1/_2$	11.4	114	28	71.1	711
5	12.7	127	29	73.7	737
6	15.2	152	30	76.2	762
7	17.8	178	31	78.7	787
8	20.3	203	32	81.3	813
9	22.9	229	33	83.8	838
10	25.4	254	34	86.4	864
11	27.9	279	35	88.9	889
			36	91.4	914

INDEX

A

Abrasives:
    for gloss adjustment, 188
    super-fine, 188
    types of, 7, 10, 11
Absorption, decreasing, 19
Adhesion, increasing, 19-20
Alder, finish for, 156-61
Ammonia, fuming with,
    56-60
Antiques, finish for, 39, 40
Arrises, sanding, 10

B

Birch:
    aging, 164
    finish for, 156-61
    fumed, 60
Bleaches, 61:
    for aging, 165
    physics of, 64
    for pickled finish, 62
Blotchiness, avoiding, 18,
    29, 82
Brushes:
    buying, 73
    cleaning, 72
    natural-bristle, 68, 70
    quality, 68, 71
    for shellac-dye mix, 180
    synthetic, 70-72
    for varnish, 97, 98-99
    for waterborne finishes,
      151
Burn-in sticks:
    applying, 184-85
    flexible low-temperature,
      183
    gloss of, 187-88
    leveling, 183, 185,
      186-87
    masking for, 184-85
    for repairs, 181-88
Butternut, fumed, 60

C

Carvings:
    rubbing out, 105
    sanding, 10-11
Cherry:
    darkening, 158-59, 164
    finish for, 156-61
    fumed, 60
Chlorine, action of, 63
Cleaning, of old finishes,
    108, 110
Cocobolo, sealers for, 19
Colors:
    artists', 35, 39-40
    for fillers, 13
    Japan, 41

matching, 55, 180
    mixing, 34-35
    for putty, 24-25
    for water-borne finish,
      143
Curves, sanding, 10

D

Distressing, process of, 165,
    168
Dyes:
    alcohol-soluble, 31, 45,
      46, 48, 179-80
    amber, 160, 161, 164
      for aging cherry, 161
    "aniline," 179
    bleaching out, 63
    defined, 52
    fading of, 45
    gel, 30-31
    India ink for, 48
    layering, 48
    mixing, 47-48
    nature of, 29-32
    non-grain-raising, 30,
      46-47, 48
    oil-soluble, 30, 45, 46, 48
    properties of, 163-64
    removing, 48
    for water-borne finish,
      143
    water-soluble, 30, 42-43,
      48
    See also Stains.

E

Ebony:
    bleaching, 62
    from walnut, 48
Edges, sanding, 10
Eye protection, for ammonia,
    58

F

Faux finishing: *See* Glazing.
Fiberboard, edge voids in,
    175
Fillers:
    applying, 15-16
    auto body, 174-75
    bases for, 13, 14
    compatibilities of, 20
    drying time for, 16
    making, 101
    paste, 94
    putty for, 175-76
    sanding, 16
    spackle as, 175
    and stains, 14
    tinting, 13-14
    woods for, 13

Finishes:
    aged, 162-66
    cleaning, 108
    compatibilities of, 20, 32
    exterior greyed, 64
    layering, 20, 48, 50-55
    matching, 180, 182-84
    nontoxic, 189-94
    See also Repairs. *separate*
      types.
French polish. *See* Lacquer,
    padding. Shellac,
    padding on.
Fuming:
    with ammonia, 56-60
    finish over, 60
    and tannic-acid
      treatment, 60

G

Glazing:
    for aging, 165-66
    process of, 50-51, 53-54
Gloss. *See* Sheen.
Glues:
    hide, 45, 170
    putty from, 24-25
Gluing up, and sanding
      sequence, 9
Graining:
    of finish repairs, 183, 187
    See also Glazing.
Grain raising, process of, 44,
    81-82

L

Lacquer:
    abrading, 11
    aerosol, 187-88
    coats for, 103
    cure time for, 103
    over finish repairs, 183
    padding, 91-92, 92-95,
      94, 180
    as sealer, 41, 53, 169
    spray, 41, 51, 52-53
    surface preparation for, 9
    See also Water-borne
      finishes.
Light fixtures, for natural
      color, 55

M

Mahogany:
    aging, 164
    bleaching, 62
    finish for, 20, 48, 75
    lightening, 164-65
Maple:
    aging, 164
    finish for, 156-61
    fumed, 60

Marbleizing. *See* Glazing.
Moldings:
    rubbing out, 105
    sanding, 10

N

Nylon pads, grades of, 8

O

Oak:
    aging, 164
    bleaching, 65
    fuming, 56-60
Oil:
    abrading, 11
    beading with, 83-84
    Danish, 79
    over fuming, 60
    linseed mixture, 76-78
    maintenance of, 78-79
    mineral, 79, 192-93, 194
    mixed, toxicity of, 189-91
    nondarkening, 79
    repairs to, 78-79
    shellac over, 158, 161
    surface preparation for, 9,
      74-78
    30-day, 74-79
    tung, 79
    tung-alkyd, 79
    two-day, 80-84
    vegetable nontoxic,
      189-91, 193
Oxalic acid, action of, 62,
    64-65

P

Paint:
    antique finish with,
      167-71
    automotive, sprayed,
      172-77
    planning for, 173-74
    primer, 175
    sealers under, 19
Patina, creating, 162-66
Peroxide, action of, 61-62
Pickled finish, bleaching, 62
Pigments:
    defined, 52
    inorganic, 64
    for water-borne finish,
      143
    See also Colors.
Pigment stains. *See* Stains.
Pine, aging, 162-65
Planes:
    for aged look, 162-63
    for flattening, 9

Polish:
    abrasives for, 104-105
    automotive, 11
    felt for, 11
    rottenstone, 11
    steel wool for, 104
    on unfinished wood, 11
    See also Rubbing out.
        Wax.
Polyurethane finish, cure
        time for, 103
Potassium dichromate,
        toxicity of, 32
Pre-stain conditioner. See
        Sealers.
Putty:
    auto, 24, 175-76
    epoxy as, 24-25
    hiding, 22-23
    lacquer-based, 23
    latex, 23-24
    making, 25
    oil-based, 24

R

Repairs:
    with burn-in sticks,
        181-88
    with shellac-dye mix,
        179-80
Respirators, for ammonia,
        57-58
Rosewood:
    lightening, 164-65
    sealers for, 19
Rubbing out:
    coats for, 103
    compounds for, 101, 177
    of intricate surfaces, 105
    of painted finish, 177
    of shellac, 90
    steel wool for, 90
    of varnish, 100-101

S

Sanders, 8-9, 18
Sanding:
    abrasives for, 7
    and blotches, 160
    cleaning after, 11
    of finish, 103-104
    grips for, 10
    by hand, 10
    machine, 8-9
    materials for, 7
    for oil finishes, 9
    papers for, 7-8
    process of, 81-83
    sealers and, 18
    sequence for, 6-7
    stearated, 8, 147-49
Sanding blocks, using, 10

Sealers:
    for blotch avoidance, 18,
        29, 45
    compatibilities of, 19, 20
    effects of, 17-20
    over fillers, 16
    lacquer as, 41, 53, 169
    oil for, 87, 88
    for paint, 19
    sanding, 18
    shellac as, 17, 18, 19-20,
        41, 46, 53
    shopmade, 29
    over stain, 14, 29, 41
    types of, 18
    vinyl, 19
Shading, defined, 48
Sheen, measuring, 106
Shellac:
    cloth for, 87
    coats for, 103
    cure time for, 103
    and dye mix, 179-80
    maintenance of, 90
    as nontoxic finish,
        193-94
    over oil, 158, 161
    padding on, 88-90
    properties of, 85-87
    putty from, 25
    as sealer, 17, 18, 19-20,
        41, 46, 53
    test for, 87
    See also Burn-in sticks.
Spackle, as filler, 175
Spray equipment:
    airless, 121
    brushing finishes with,
        149
    fluid tips for, 131
    high-pressure, 119-20
    HVLP, 119-20, 122-27,
        123-25, 124, 126,
        131
    lazy Susan, 176
    operating, 134-35
    setting, 131-32
    units for, 115, 116-19
    for waterborne finish,
        152
Spray finish:
    benefits of, 114-16
    booths for, 129
    compatibilities of, 149
    finish straining for, 130
    humidity for, 130, 152
    problems with, 134-35,
        149
    sequence for, 132-33
    surface preparation for,
        129
    temperature for, 130, 152
    viscosity for, 129-30

Stains:
    applying, 30-31, 32
    brushes for, 34, 54
    chemical, 32
    color sticks for, 55
    dry brushing, 33-37,
        53-54, 166
    and fillers, 14
    lacquer over, 37
    layering, 32
    mixing, 32
    nature of, 28
    oil-based, 34-35, 38-41,
        40
    putty under, 25
    sealers for, 14, 29, 41
    shading, 53
    surface preparation for,
        30
    testing, 31
    See also Colors. Dyes.
Stains (blemishes), bleaching
        out, 64-65
Steel wool:
    cleaning with, 108
    grades of, 8
Stock preparation, sequence
        for, 9
Surface preparation:
    for aged look, 162-66
    for antique painted
        finish, 168-69
    for fillers, 13
    for irregular pieces, 10
    for lacquer, 92
    for oil finish, 75-78
    for paint, 174-76
    for shellac, 87-88
    for spray finish, 129
    for stains, 30
    waiting time after, 78

T

Tannic acid, bleaching out,
        64-65
Teak:
    lightening, 164-65
    sealers for, 19
Toning:
    mixes for, 55
    process of, 48, 51, 52-53,
        54-55
Turnings:
    lacquer for, 95
    rubbing out, 105
    sanding, 10

V

Varnish:
    abrading, 11
    applying, 98-100
    brushes for, 97, 98-99
    cure time for, 103

    fish-eye preventative for,
        99
    gloss of, 104-105
    properties of, 97-98
    quality, 98
    surface preparation for, 9

W

Walnut:
    aging, 164
    ebony simulation with,
        48
    lightening, 62, 63,
        164-65
Water-borne finishes:
    brands of, 138-45
    brushes for, 151
    brushing on, 146-52
    coats for, 103
    color of, 143
    cure time for, 103
    flow additives for, 151
    grain raising for, 143, 147
    heating, 152
    spraying, 152-53
    topcoats of, tinting, 143
Wax:
    bees-, and mineral oil,
        194
    buffing, 109
    clear, over tinted, 110-11
    historical use of, 107
    liquid, 171
    as nontoxic finish, 193
    over oil finish, 84
    remover for, 108
    tinted,
        for aging, 166, 171
        brands of, 108, 109
        as finish repair, 108
Wood:
    grain of, mimicking, 23
    matching, 34, 62
    oily, sealers for, 19
    refinished, bleach for, 64

Publisher: Jim Childs

Associate Publisher: Helen Albert

Associate Editor: Strother Purdy

Indexer: Harriet Hodges

Designer: Amy Russo

Layout Artist: Lori Wendin

Fine Woodworking magazine

Editor: Tim D. Schreiner

Art Director: Bob Goodfellow

Managing Editor: Jefferson Kolle

Senior Editors: Jonathan Binzen, Anatole Burkin

Associate Editor: William Duckworth

Assistant Editor: Matthew Teague

Associate Art Director: Michael Pekovich